CHIMPANZEES

CHIMPANZEES

Tess Lemmon
WITH A FOREWORD BY
Jane Goodall
ILLUSTRATIONS BY ROBIN BUDDEN

Whittet Books

TITLE PAGE ILLUSTRATION: family group.

Dedication
For Mith and Jerry

First published 1994
Text © 1994 by Tess Lemmon
Illustrations © 1994 by Robin Budden
Whittet Books Ltd, 18 Anley Rd, London W14 0BY
The right of Tess Lemmon to be identified as the author of this work has been
asserted in accordance with the Copyright, Designs and Patents Act 1988
All rights reserved
Design by Richard Kelly
British Library Cataloguing-in-Publication Data. A catalogue record for this book
is available from the British Library.
ISBN 1 873580 04 5

Typeset by Spectrum City
Printed and bound in Great Britain by
Biddles Ltd, Guildford and King's Lynn

CONTENTS

Chimp handprint, reduced by a third.

Foreword by Jane Goodall

It was three decades ago, in 1960, that I began to study a community of chimpanzees living on the eastern shores of Lake Tanganyika in Tanzania (it was Tanganyika then). That research is still going on with data collected daily by a dedicated team of Tanzanian field staff and foreign researchers and students. Since 1960 other scientists have observed chimpanzees in other parts of their range in Africa so that, today, we know a lot more about these apes than we did thirty odd years ago. But there is still a great deal more to learn, most particularly about the differences in behaviour that have been recorded in different populations. These can be described as cultural traditions, passed from one generation to the next through observational learning.

There are, today, a number of books written about chimpanzees. What makes this one special is the person who wrote it. Tess Lemmon was fascinated by monkeys and apes as a very small child and always chose monkey toys rather than dolls. As she grew older, she nurtured that childhood passion and wove it into her career; in this regard I feel a certain kinship with her since I did much the same. During her life Tess made major contributions to the understanding of monkeys and apes – because she not only cared, but cared enough actually to work with them. It was as a result of the time she spent with orphan chimpanzees in two sanctuaries in Africa that she learned to understand them so well. After all, her teachers were the best possible – the chimpanzees themselves.

Tess was thus well qualified to write this book. She describes the life of chimpanzees in the African forests: their food preferences and foraging behaviour, hunting techniques and tool-using and -making skills. And she writes with real understanding about their social behaviour: their family life, dominance hierarchies and complex communication patterns. Chimpanzees are our closest living relatives in the animal kingdom, and the many ways in which their behaviour resembles our own become very clear as the book unfolds.

Here Tess has distilled the essence of what a number of scientists have learned about chimpanzees. She has written clearly and simply to help give a better understanding of the true nature of chimpanzees, with a sensitivity and understanding gained from her own experience.

Chimpanzees are fast disappearing in Africa, partly because of habitat destruction, partly because they are sought after for bush meat, and partly as a result of a demand for live infants – for pets, for entertainment, for biomedical research. There is much cruelty involved, not only during the hunting and export (which is mostly illegal), but also inflicted by the conditions of captivity all over the world. Tess deals with these issues, not in an over-emotional, irrational manner, but factually. Again, much of her knowledge was derived from her own first-hand experience.

Tess's untimely death has deprived the chimpanzees of a true friend, but this book will serve as her legacy to them for all time. And thanks to the understanding of her mother, Tess's royalties are to be divided between two organizations that work to protect chimpanzees and give them a better life. So everyone who buys this book is helping them too. Thank you.

Jane Goodall
July 1993

Introduction

About one hundred years ago some chimpanzees in Gabon came across a strange contraption in the middle of the forest. Most of them were very wary of it and kept their distance, a few plucked up courage to dart past it, and one or two edged up and took a good look. It was a cage, and it had a man inside it. This exotic exhibit was there not for the chimpanzees to look at the man, but for the man to look at the chimpanzees. His name was R. L. Garner, and he was the first fieldworker to go into the forest to study chimpanzees. Fearing attack from ferocious wild beasts, he built the cage for his protection and locked himself inside it every night and most of the day for almost four months.

But he did not emerge with much to show for it, because the chimps did not often drop by and behave in front of him. They preferred to be heard, not seen, and Garner went home with vivid accounts of their screams and shouts and their drumming, which he thought they performed on drums they made from clay.

Since Garner, chimpanzees in various parts of Africa have got used to assorted tents, cabins and huts, and to people emerging from them. Indeed, some chimpanzees might find it odd not to have a human tagging along! Fieldwork really took off in the 1960s when Adriaan Kortlandt, Vernon Reynolds, Kinji Imanishi, Junichero Itani and Jane Goodall each ventured into chimpanzee territory, armed with questions, most of which had more to with humans than chimpanzees. It was thought that, being our closest living relatives, chimpanzees might be able to put flesh on the bones of our ancestors and help us see our past more clearly.

They did (Jane Goodall has described watching chimpanzees as being like turning back the clock); but they also intrigued their observers so much that the search for our roots soon grew into interest in them for their own sakes – and continues to grow. Thanks to over thirty years of observation, Jane Goodall has made the chimpanzees of Gombe the stars of the longest-running fieldwork project ever undertaken. One hundred miles to the south of what is now the Gombe National Park in Tanzania lies the Mahale Mountains National Park, where Japanese primatologists led by Toshisada Nishida have been studying chimpanzees since 1965. Over the years many other observers in different places have added to the knowledge and insights of fieldworkers at Gombe and Mahale, and several long-term projects are in full swing: Christophe Boesch and Hedwige Boesch-Achermann in the Tai Forest of the Ivory Coast, Caroline Tutin and Michel Fernandez in the Lope Reserve of Gabon, Richard Wrangham and Isabirye Basuta in the Kibale Forest of Uganda. This book draws on their work and on that of others too numerous to mention.

Chimpanzees are some of the best-studied animals in the world - and the more we learn the more there is to learn. Some of the most intriguing news now emerging from the field is how much they vary from group to group and individual to individual. Like people, they have different habits, family traditions and local customs; there are quiet chimps, noisy chimps, grouchy chimps, easy-going chimps. It is as difficult to generalize about them as it is to generalize about people. This book is peppered with 'usuallies' and 'oftens' and seemingly straightforward questions such as, Are they territorial?, Nice to their children? Aggressive? are answered by 'Yes and no', 'It depends'.

Jenny, an adult who lives in a forest in West Africa, was the first chimpanzee I ever met. After we had ignored each other for a while, she took my hand, led me to a fallen tree and sat me down beside her. Then she peered into my face and, like a brisk hairdresser, put her hand firmly on my head, turned it on one side and proceeded to groom my eyebrow. The long, scratchy nail of her forefinger worked its way from my eyebrow to the corner of my eye and up to my hairline before she stopped and looked me over, then released her grip on my head, turned away, wiped her nose with the back of her hand and waited. Finding me slow on the uptake she turned to face me again and lifted her bristly chin. Imitating her grooming technique as best I could, I tweaked and parted the hair on her neck.

Chimpanzees groom to keep each other clean, but it is also the most important way of demonstrating good feelings towards someone else. Trust, reassurance, respect and friendship are all tied up with grooming. Thinking about it afterwards, I was struck by Jenny's size and strength – and by her presence. I found it unnerving and exciting to be told what to do by her, and to be looked at with such understanding, with eyes that held mine.

Above all, I felt honoured. As a member of the Chimpanzee Rehabilitation Project, Jenny had been rescued from a pet shop and had needed time to adjust to living wild again, but when I met her she was perfectly at home in the forest and had no need of humans. I had nothing to offer her, yet she and other rehabilitated chimps allowed me a glimpse into their world, and the more I got to know them the more my respect for them deepened. Being with them in the forest convinced me that no chimpanzees, for whatever reason, should be robbed of their own way of life – their own natural dignity – and be held captive. Yet only one hundred years since Garner cleared himself a space and put up his cage, we have bulldozed our way into their world and captured them to be our playthings, clowns and laboratory guinea pigs, and we have demolished their homes to make room for ourselves, shoving them to the brink of extinction.

This book aims to show how chimpanzees live in the wild. It also looks at the mixed feelings we have about our closest relatives, at what a raw deal they have had from us, and at what is being done to stop them from dying out completely. Thanks goes to all fieldworkers who have contributed to the understanding of chimpanzees, and to Stella Brewer, Janis Carter and David and Sheila Siddle for giving me the opportunities to get to know some chimpanzees and to watch them with each other in the forest. Special thanks to Carole Noon for helpful discussions and encouragement from the earliest stages of this book, to Geza Teleki, chairman of the Committee For The Conservation And Care Of Chimpanzees, for information and interest, and to Annabel Whittet for patience and enthusiasm. The biggest thank you of all goes to Jane Goodall who I have looked up to for as long as I can remember and who has done me the honour of reading the manuscript of this book.

Tess Lemmon
1992

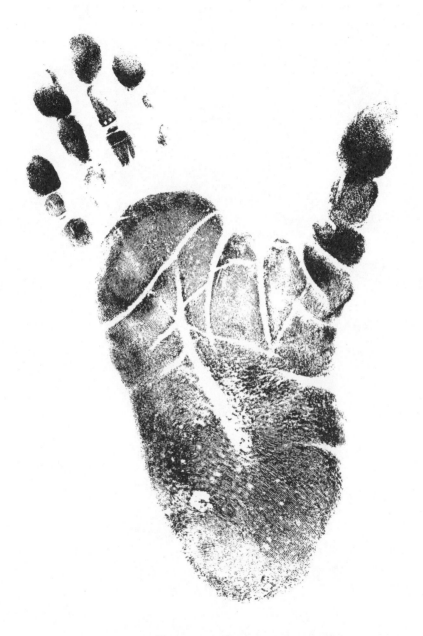

Chimp footprint, reduced by a third.

• 1 •
Fingers and Thumbs

tupaia

lemur

monkey

chimpanzee

Everyone knows what chimpanzees look like. But how and why did these black-haired, gangly animals get here? Why the squashed noses and big ears? Why do they walk the way they walk? What *is* a chimpanzee?

A chimpanzee is a primate, along with about 200 other animals, from tiny monkeys that could climb up your finger to big gorillas weighing more than two men. They are strikingly varied not only in weights and measures but also in looks and lifestyles. They come in assorted colours from bright orange to jet black, and sport a range of hairstyles, beards and moustaches. Some live alone, others in pairs, others in big groups; some eat only insects, others mainly leaves, others pick and choose their way through a selection of meat and veg. If you watched a chimpanzee and a bushbaby each getting their first meals of the day, they would not seem to have much in common. The chimpanzee wakes up hungry at dawn. It climbs down from its nest in a tree and sets off along familiar trails through the forest. From some distance away it spots a cluster of ripe figs high in a tree. The chimp clambers up the tree, then stops halfway and peers through the leaves at the figs hanging from the tip of a slender branch. It climbs higher, picking its way more carefully now among the thin branches. It spreads its weight between them, and sometimes swings its body below them, holding on with one arm then the other. By dangling like this it manages to reach the figs. It stretches its arm, snaps several from the bunch, drops down to a sturdier branch – and tucks in.

The bushbaby gets up for breakfast in the middle of the night. After a long day's sleep it climbs out of a hole in a tree and sets off along a branch, running at top speed on all fours. As if following paths along the ground, it weaves its way through the tree – and suddenly leaps into mid-air, bushy tail flailing. Feet and hands land smoothly on a branch below, and it stands stock still, its big round eyes and big ears taking in the sights and sounds of the night. Then, quick as a flash, it darts out an arm, plucks a passing moth from the air – and tucks in.

The chimpanzee eats a wide variety of fruit, leaves, nuts, buds, meat and eggs. It finds some food on the ground, some in the trees, and eats mainly in the early morning and late afternoon. Meals can be noisy affairs, when family and friends get together. The nocturnal bushbaby eats in solitary silence, and mostly sticks to insects. There are shades of the stealthy hunter in the way it watches, waits and strikes.

Yet, for all their differences, the two primates are basically alike. Both are not just at home in the trees, they are at home in a similar way. Both climb with great skill, holding on with hands and feet, and instead of putting their mouths to their food like most animals, both take their food to their mouths. Climbing trees, picking figs and catching insects are not as simple as they look. They are actually quite complicated procedures that require just the right equipment. All animals are shaped by their environment, and for 65 million years evolution has been whittling away at primates to make them fit the trees.

It picked an unlikely candidate to work on. Snuffling through the leaf litter on the floor of a forest in south-east Asia is an animal that goes by the name of tupaia. It looks like a cross between a squirrel and a rat with an elongated nose, and it is not the sort of animal to attract much attention. But, as far as taxonomists are concerned, it is one of the most exciting creatures in the universe. They have decided that it resembles some of the earliest mammals from which many different groups – including primates – evolved.

Go back 100 million years or so, and you would have found an animal very like today's tupaia dodging the dinosaurs' feet. No one really knows what upset the dinosaurs and made them go, but their departure left a lot of gaps. There was food to eat and places to live, and the tupaia-like animals seized their chance. From them various groups of mammals evolved. Some took to the trees, which were no longer being browsed by huge vegetarians.

Today's tupaias sometimes climb trees to eat insects and seeds (they are more commonly known as tree shrews, which is misleading as they are neither shrews nor tree-dwellers). Like a squirrel, a tupaia in a tree runs along the branches, digging its claws in as best it can. Imagine how much easier it is for a bushbaby or a chimpanzee, with fingers and toes that wrap *around* the branches; how much easier it is to get a grip if instead of ending in brittle points, you have sensitive, fleshy pads protected by flat nails; how much easier it is to feel things and hold them with a hand. Most mammals have paws, hooves or flippers that are designed to do only one or two things. The hand – with its palm, fingers and thumb – is a multi-purpose gadget. It is strong and delicate, and the whole of it can be used, or just a part.

Being able to make the fingers work on their own or bring them together gives a lot of scope for probing, feeling and holding; add to this a thumb that goes off at an angle, an 'opposable' thumb, and you have something that can meet a fingertip in the most precise pinch, or stretch away from the fingers to make the hand into the firmest clamp. When the hands are busy reaching and picking, the feet are needed for holding on. They can do so much better if they, too, can clamp. Primates have opposable big toes, so their feet look like hands.

Swapping paws and claws for fingers and toes with nails gave primates agility, strength and dexterity, and helped them take over the trees. But other developments also took place. A tree is a multi-level maze. It twists and turns, and its thick leaves cut out the light, hiding abrupt endings and sudden drops. When you are in a tree you need to see where you are going; you need your eyes to be good, and in the right place. The tupaia's eyes are on the side of its head so they look in different directions and the tupaia follows its nose, zigzagging through the mouldy leaves on the forest floor. In this dark, flat world, smell is far more valuable than sight.

In its world of many layers way above the ground, the primate relies on its eyes. They are in the front of its face, and both focus on the same thing. This is called 'binocular vision', and it gives an accurate, 3-D view of what is ahead. Primates can not only judge distances, but also find fruit hidden in the foliage. And being able to see in colour, they can tell just by looking at it whether it is green and hard or yellow, red, or purple and ready to eat.

As the primates began to take shape, so too did their control centres. To be able to spot something and reach out and get it they needed good hand-eye co-ordination, and for that they needed good brains. Much of the brain's expansion was devoted to working the hands and eyes – but it was a two-way process: like a computer, the brain is at the mercy of what it is fed, and the more the quick-witted primates took in, the bigger their brains grew.

Most animals use one sense more than the others, but primates have a good sense of smell and hearing as well as excellent eyesight. Even so, with the rise of sight over smell, the primate nose shrank.

Shrinking noses, shifting eyes, and developing digits are easy enough to talk about but hard to imagine. Like the tupaia, there are animals alive today that give some idea of how the changes happened.

The one that looks most as if it is on the way to becoming a monkey is the lemur, and it belongs to the sub-order of primates known as prosimians – which literally means pre-monkey. There are 35 species of prosimian, including bushbabies, lorises, and over 20 different lemurs. One of the best known is the ringtail, which is about the size of a cat. Its nose is much shorter than the tupaia's but is nevertheless substantial – and wet – showing that smell is important. Its eyes face forwards but their fields of vision do not completely overlap, so it cannot see in much depth. It has fingers and toes, but they tend to work together, so dexterity is limited. It eats fruit and leaves, and, as if halfway towards picking its food with its hands, it pulls whole branches to its mouth.

About 50 million years ago lemur-like animals and other prosimians were widespread, but during the next 10 to 20 million years more agile,

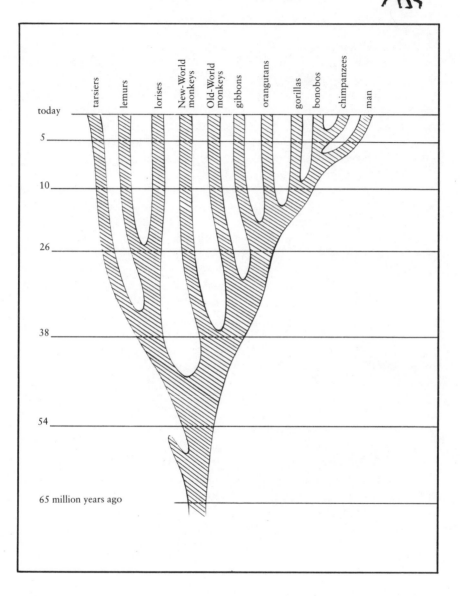

Primate evolution.

brainier monkey-like animals sprang up, and were too much competition for the prosimians, so many of these died out. The only place you can find lemurs today is on the island of Madagascar where they have flourished because there have never been any monkeys there.

Lorises and bushbabies found a way of living right under the monkeys' noses without having to compete with them face to face, and the arrangement is still working today: they come out at night. In Africa, for instance, while monkeys sleep, bushbabies take their turn in the trees, using speed to save themselves from the jaws of cats and snakes.

By 30 million years ago, monkey-like creatures had spread across Africa, Asia and America. Today there are over 130 species of them: the Old World monkeys of Africa and Asia, and the New World monkeys of Central and South America.

You can tell just by looking at them that they are tailor-made for the trees. Their long arms and legs and lithe bodies are built for running and jumping, and they are saved from falling by their hands, feet and tails. Held straight out behind when running, the tail acts as a balance, and as a rudder in mid-leap – and some are even like an extra arm, with a palm on the end that grasps branches.

Yet for all its attributes, a monkey has diverged very little from the basic model of a four-footed mammal: long, narrow trunk supported by four limbs. Monkeys sit up, but mostly they stand and walk on all fours – which works very well in the trees up to a certain size, but evolution tended to make primates get bigger and bigger. Some monkeys weigh eight times as much as a tupaia. While placement and shape of some features seem obvious when you think about them, evolution's reasons are not always clear-cut, and experts are unsure why primates grew. Some suggest it is all to do with male rivalry: that when they showed off in front of each other and tried to impress females, the bigger ones did better. Others suggest it was a combination of factors to do with diet, reproduction and standing up to predators. But the bigger you are the harder it is to keep your balance on top of a branch, so the larger primates move through the trees in quite a different way. Instead of holding their bodies horizontally and walking along the branches, they hold their bodies upright and hang by their arms beneath the branches. These are the apes.

Gibbons, siamangs, orangutans, gorillas, bonobos and chimpanzees are all apes. Gibbons and siamangs are known as 'lesser apes' because they are smaller than the others which are all 'great apes'. The easiest way to tell an ape from a monkey is that apes do not have tails: they are no longer needed. Also, an ape's arms are longer than its legs. Early apes developed long, strong arms attached differently at the shoulders to give them greater

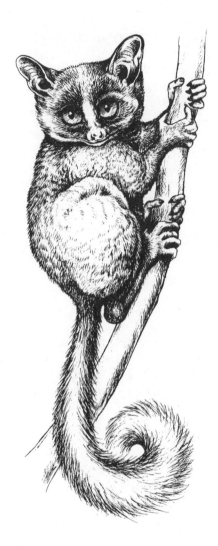

Bushbaby.

mobility. Monkeys, like dogs, for instance, cannot extend their forelimbs sideways, whereas apes can.

Gibbons and siamangs are the acrobats of the primate world. By holding on with one arm then the other, they swing through the trees at top speed. Their ancestors were probably larger, but by becoming lighter and more nimble than many monkeys, today's lesser apes have been able to take over the treetops of south-east Asia where they dangle from the tips of slender branches to pick ripe fruit, reaching parts of trees that other primates cannot reach.

Orangutans have long reaches, but they are far too heavy to swing about. Weighing up to 90 kg (198 lb), orangs are the world's heaviest tree-living mammals. They climb slowly and carefully, often with all four limbs stretched in different directions to spread their weight.

The biggest ape gave up and climbed down. Weighing as much as 180 kg (396 lb), the gorilla is too heavy for the trees, so it lives on the ground. But it is not the only one. Some African monkeys, such as baboons and patas, also spend most of their time on the ground. Their ancestors probably came down because their increasing size made them clumsy in the trees (the biggest South American monkeys managed to stay put by developing long, strong, grasping tails). Once down, the monkeys found plenty to eat, and several of today's species do very well out of the trees. Most primates eat a variety of vegetation and insects, and the ground-dwellers find rich pickings. Gorillas munch through 20-30 kg (44-66 lb) of greens per day, while monkeys eat shoots, seeds and roots.

It is hard to imagine a koala or a sloth out of the trees, but primates are far less tied to an arboreal life than they are. Evolution spent 65 million years coming up with the perfect tree-dwellers only to find that these eyes, noses, ears and limbs could be put to good use on the ground.

The gorilla and the baboon only needed minor modifications – mostly to do with teeth and jaws so they could chew their high-fibre food. It is the same with all the monkeys and apes: all are variations on a theme, and it is their lack of specialization that has made them so adaptable and successful. The chimpanzee is probably the best all-rounder of them all. An adult female chimpanzee weighs about 30 kg (66 lb) and has a head and body length of 70-85 cm (28-33 in). An adult male weighs up to 50 kg (110 lb) and measures 75-92 cm (30-36 in). Bigger than gibbons, smaller than gorillas and orangs, chimpanzees are equally at home in the trees and on the ground.

In trees they spread their weight like orangs, and also swing like gibbons with their long, strong arms and long fingers that hook over the branches. On the ground they walk on all fours like gorillas. The front part of the body is propped up on the long arms and the back slopes down to the shorter legs. The hands are turned into feet by resting the weight on the knuckles of the folded fingers. Known as 'knuckle-walking', it is a rather ungainly gait, but efficient enough; a chimp can turn its gangly stride into what has been described as a 'rapid gallop'. Chimps also walk upright, though only a few paces at a time. Upright, they are bow-legged and flat-

Chimp knuckle walking.

footed, and their arms come to below their knees.

One reason for standing up is to free the hands, and chimps cup their fingers and palms to carry food. Having such long fingers means that the thumb cannot reach the tip of the forefinger, so to hold anything small or delicate the chimp brings its thumb to the side of the finger.

At first glance the chimpanzee foot looks more like a hand, and the gap between the big toes and the other toes shows how well it can grasp a branch. While chimps use their hands to walk on, when they are sitting down their feet sometimes double up as hands to hold food.

Big, round eyes set close together give chimps excellent binocular vision. The nose is a small but useful back-up to the eyes: chimps sniff food, each other and anything unfamiliar. Inside the big mouth is a set of 32 relatively unspecialized teeth, with large, broad molars for chewing fruit, greens and meat. Chimpanzee ears have been variously described as 'huge', 'cauli-flower' and 'immense and finely-modelled'. Chimps do have good hearing. They often stop and listen to sounds around them and call to each other to keep in contact when they are out of sight in the forest, but quite why their ears are so big is a bit of a mystery.

Ticking away inside this versatile body is the big brain. Relative to their body size, apes have bigger and more convoluted brains than monkeys,

and chimps have the biggest brains of all. The bigger the brain, the less tied an animal is to fixed ways of doing things. The small-brained tupaia, koala and sloth do not have much flexibility: they lead simple lives controlled by a handful of needs, and one day is very much like another. For a chimpanzee, no two days are the same.

The chimpanzee is always having to make decisions about what to eat, where to go, what to do next. Its brain has well developed learning and memory compartments that offer it a range of options and a store of knowledge to help it choose the most appropriate one.

Chimpanzees are the most adaptable primates of all – not just in body design but in behaviour too. If no two days are the same, no two chimpanzees are the same either. Their big brains have freed them to become such individuals that experts hestitate to pronounce on what they definitely do or definitely do not do, because sooner or later one particular chimpanzee pops up and proves them wrong. As Dr Roger Fouts writes: 'The chimpanzee is a mythical beast.'

THE FOURTH GREAT APE

Bonobo, as compared with chimp (left).

Deep in the forests of Zaire live some animals that used to be regarded as just another race of chimpanzee, but in 1929 they were formally identified as a separate species, and named pygmy chimpanzees. They are more slender than chimpanzees, but hardly any smaller, and are now more commonly known as bonobos (probably a mispronunciation of Bolobo, a town on the Zaire river). Long legs and narrow shoulders make bonobos more lanky than chimps, and they have slightly smaller heads, dark faces and red lips – and the long, fine hair on their heads is neatly parted down the middle.

Bonobos live only in the dense equatorial forests of central Zaire, south of the Zaire river (which separates them from both the eastern and central races of chimpanzee). For a long time they were just mentioned in passing when talking about chimpanzees, but thanks to several long-term field studies that started in the early 1970s, interest in them is growing and whole books have been written about them.

At last a picture of the fourth great ape is emerging. Like its looks, its way of life resembles chimpanzees' in outline, but differs in important details that give it its own distinctive character. Both are group-living, fruit-eating forest-dwellers, but the more graceful and agile bonobos spend more time in the trees and are more dependent on fruit. Like chimps, they live in big groups but these are more tight-knit, and in bonobo society it is the females who form the stable core of the group, whereas with chimps it is the males.

Like chimpanzees, bonobos are endangered. Coffee and cacao plantations are replacing their forests. In some areas they are hunted for meat, and they are caught alive for the zoo and pet market. At Wamba, in Zaire, the local people are very involved in helping to study and protect the bonobos that Japanese primatologist Dr Takayoshi Kano has been observing for twenty years. In 1987, when Dr Kano was away, soldiers turned up at Wamba to capture some bonobos. The trackers who work with Dr Kano refused to lead the soldiers to the bonobos and were beaten up. After the soldiers had found and killed some bonobos, the chief tracker saved another group by barring the way with his own body, and telling the soldiers they would have to kill him first. Political unrest forced all fieldworkers to leave Zaire at the end of 1991, and no one really knows what is happening to the people, the bonobos or the forests.

• 2 •
Them and Us

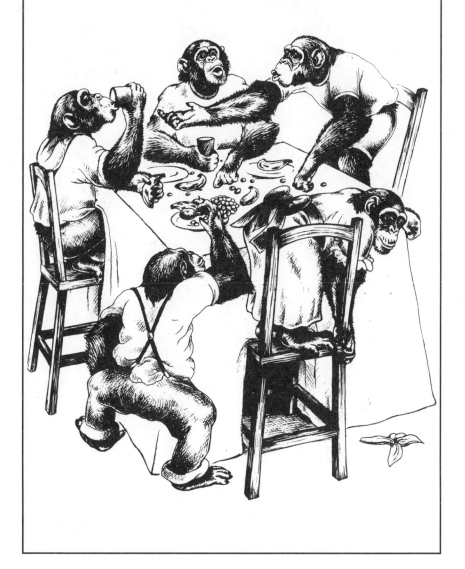

Being a cousin of the bushbaby is the not the chimpanzee's main claim to fame. It is best known for being the closest relative of a primate not yet mentioned: the human. Chimpanzees and humans share nearly 99% of their genes. Chimpanzees are probably closer to humans than they are to their fellow African great apes, gorillas. Chimps and humans are as alike as grizzly bears and polar bears.

It has only been in the last few years that the extent of our kinship has fully come to light. Until then, the four great apes (chimpanzees, bonobos, gorillas and orangutans) were thought to be closer to each other than any one of them was to humans or to lesser apes (gibbons and siamangs), and are classified accordingly. The superfamily Hominoidea, to which all apes and humans belong, is subdivided into three families: Hylobatidae (lesser apes), Pongidae (great apes) and Hominidae (humans).

By painstakingly piecing together bits of fossilized bones and teeth, paleo-primatologists have traced many primate ancestors and placed all their living descendants on a family tree. They used to picture it as more of a ladder than a tree, with a person perched at the top and all the other primates strategically placed on the lower rungs, the apes having managed to clamber the highest. It implied either that people *evolved from* the apes, or that all primates are waiting to be turned into people, with the apes at the front of the queue. Now they draw more of a bush, and put all the different primates on their own branches (see page 17).

With only fragments to go on, the experts can seldom draw direct lines between fossil and modern primates, or pinpoint exactly when new branches began. It must be like trying to do a stack of jigsaw puzzles with most of the pieces missing. There is a lot of intelligent guesswork, and a lot of arguing about what goes where. But there was general agreement that humans and each of the great apes had started their own branches by 18-14 million years ago.

More fossils, new ways of looking at them, and new, improved methods of dating them caused a re-examination of the whole picture during the 1980s, and it began to look as if apes and people had stayed together much longer than previously thought. This was confirmed by a new investigation technique: instead of mixing and matching bits of ancient bones, bits of blood and tissue are extracted from living bodies in order to compare the molecules of close relations and find out exactly what genetic material they share. One group of animals to have been scrutinized are horses and zebras, but when apes and humans were put under the microscope it was discovered firstly that the chimpanzee and the gorilla were closer to the human than either is to the orangutan, and secondly that the DNA of chimps, gorillas and humans is so similar that they must have parted company much more recently. The general feeling now is that gibbons and siamangs branched off 12 million years ago (MYA), orangutans 10 MYA, gorillas 10-8 MYA, and chimps and humans not until 5 MYA.

But there is still a lot to find out about them and us. Comparing molecules has produced the most complete story so far, but everyone is quick to say that it is only the latest way of looking for clues, and only one of many lines of inquiry. It does seem that the more we look the closer we get to the apes, and every scrap of evidence is used to try and sort out exactly when each branched off and who went first. After studying certain details of bone structure, some primatologists believe that orangutans are our next of kin, while others think that chimps must be closer to gorillas than they are to us because chimps and gorillas share 41% of parasites and chimps and people only share 33%!

For all the lopping off and grafting on, the basic shape of the tree has stayed the same – but it may no longer make sense to keep chimps and gorillas in one family and humans in another. Most re-organization plans try to find ways of allowing them to join us as a sub-section of Hominidae. Some see this as a return to the ladder, with the apes being hauled on to the top rung as honorary humans or sub-humans; others find it offensive to put them in with us. An alternative is to keep everyone apart but give them equal rank, so that *alongside* each other would be orangutans, gorillas, chimpanzees, humans.

'Thoughts about the evolutionary process', writes philosopher Mary Midgley, 'have never, from their outset, been conceived only as inquiries about the physical world. They have always been directed also towards the problem of human destiny and human uniqueness.' We have very mixed feelings about our closest relatives. Part of us reaches out to shake hands, the other part withdraws to count our differences. At the same time as threatening our identity, they help us find it, for we cannot really decide what makes us special unless we have someone to measure ourselves against. We rub shoulders and welcome them for bridging the gap between us and the animal world, but worry if they get too close.

Another philosopher, Peter Singer, is taking the bull by the horns by calling for chimpanzees and gorillas to be embraced within 'the community of equals' and accorded the same rights as human beings. To critics who ask why we should be especially nice to them just because they are so like us, he replies that other animals might be just as worthy but you cannot embrace them all at once. You have to start somewhere, and these particular beasts are the obvious ones to make 'a small breach in the species barrier'.

Chimpanzees often breach it at our invitation: we have many ways of making them human. They have served us as spacemen, drugs testers, substitute children and even factory workers (paid in bananas). But mostly we turn them into clowns. And the more they are made to look and behave like clumsy, stupid people, the further away from you and me they seem. These caricatures that bolster our sense of superiority are still performing tricks in circuses and nightclubs all over the world, and they are *still* the chimpanzees you are most likely to come across – on cards, posters, TV adverts. Tea parties at London Zoo were stopped in 1972, and the PG Tips chimps may soon be history now that people are pointing out how demeaning to them it is to make them behave like us.

Without its clothes and props, without its disguise, the chimp stranded in our world can seem too close for comfort. Does all the laughing and jeering at the zoo help to relieve the awkwardness and put the barrier between them and us? And all the disgust? Their bare bottoms and the way they pick their noses. Psychologist Paul Shepard writes: 'It needles us because the ape is uncomfortably close as an animal and disgustingly far away as a human.'

It has been needling us, and we have had mixed feelings about it, for a long time. Aristotle was one of the first to try and spell out how man was at the top (woman was a 'natural deformity') and all the animals were down below – but monkeys and apes have never stayed put. People in different times and places have liked and respected them for their kinship, and even made them gods. But Christian tradition has despised and feared them, and tried to keep them at bay. Shepard points out that by the time Darwin came up with the hard evidence, 'it is clear that we already feared the worst'.

In fables, sayings and paintings, the ape or the half-ape half-man is a figure of evil, vice and hypocrisy. And the ancestors of today's clowns were the man-ape fools and pranksters and the fallen or failed people, mocked and pitied. In modern stories and films, mad scientists continue creating, accidentally or on purpose, these ambiguous creatures that grab our imagination and help us think about ourselves.

Real life chimpanzees help us think about ourselves too, thanks to the humans who have climbed the barrier and gone to meet them on their own ground. Once it was a case of trying to identify individuals and find out if they lived in groups or pairs, and what they did all day. Now people are compiling detailed dossiers on them from the moment they are born, and asking how their personalities develop, and what sort of family traditions they inherit.

Detailed comparisons of populations are just beginning. The chimpanzees we know most about, at Gombe and Mahale, live in mixed savannah-woodland at the eastern edge of the chimpanzee distribution map. We are starting to discover what they do and do not have in common, and how they differ from forest-dwellers such as the ones at Tai and Lope, and why. How much is due to their habitat, and how much is habit? Each population appears to have its own distinctive variations of behaviour and as the studies continue the fieldworkers begin to seem more like anthropologists, each describing a particular society with its own customs. And one of the biggest questions they are asking each other is whether chimpanzees can be said to have a culture.

It is the biggest threat to human uniqueness so far, and it is making us take a long hard look at that part of our big brains that deals with the most advanced processes such as imagination and creativity. Wild chimpanzees keep surprising observers not just with their braininess but also the subtlety that goes with it, and now a lot of attention is being paid to what exactly goes on inside their heads. Can they think abstract thoughts? How far ahead can they plan? How good are their memories? How self-conscious are they (and therefore how aware of the consequences of their own actions)? Just how clever are they?

Some of the most valuable research is looking at why chimpanzees evolved their higher faculties, what use they are in day-to-day life, but some psychologists think the way into a chimpanzee's mind is to set it intelligence tests in a 'controlled environment' away from all the diversions of everyday life, so several universities (mostly in the US and Japan) include a number of chimpanzee students in their ivory towers (where their accommodation varies from unimaginative cages to large enclosures). The psychologists not only watch how the chimps help themselves to food

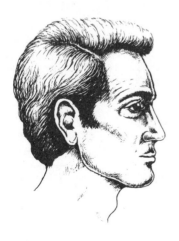

by working out the idiosyncrasies of various glorified vending machines, but also try and set up a dialogue with them.

The first attempt to break the language barrier was made in the 1950s, when Keith and Cathy Hayes tried to teach their pet chimp to talk. They did not get very far because chimpanzees do not have the physical equipment to speak, but in the mid-1960s Allen and Beatrix Gardner followed up an idea first mooted forty years earlier by primatologist Robert Yerkes, who suggested a chimp could be taught to 'talk' by using its fingers. The Gardners taught their chimps American sign language, and since then a succession of Dr Dolittles have taught a succession of chimpanzees to communicate by means of a succession of ingenious gadgets including plastic symbols and computer graphics.

The first sign-language chimpanzees lived with their teachers as part of the family, and lessons were so informal that the teachers were accused of jumping to conclusions and even of putting words into the chimps' mouths (hands)? Now the whole business is much more formal and scientific and grandly entitled 'cognition'. The chimps also appear to have progressed! Early work concentrated on the number of signs they could learn, and how they could put signs together to make new words. When she first saw a swan, for example, the Gardners' star pupil, Washoe, described it as a 'water bird'. Another chimp, Moja, described Alka Seltzer as 'listen drink'. And when at a loss for words they sometimes invented a sign. When Lucy (who was taught by Maurice and Jane Temerlin) wanted her lead put on to go for a walk, she held her crooked index finger to the ring on the collar she always wore. The humans copied her and the sign became part of the vocabulary.

Now the chimpanzees seem to be constructing sentences and the psychologists are wondering whether they know about grammar, syntax, *language*? But sometimes they seem so caught up in the details of their experiments that it is not always clear what they are learning about chimpanzees by teaching them our communication system, and treating them like children. One big reason for creating this sub-human race of articulate apes is to find out what they can tell us not about them but about us: about the origins of language. Dr Roger Fouts, who has a long and distinguished track record in this field, reminds his colleagues that chimpanzees have minds of their own ('chimpanzee intelligence is whatever chimpanzees do', he writes) and unless the experiments give them room to be themselves we could 'end up finding out more about the mental capacity of the scientist than about the ape he or she claims to be studying'. And Adrian Desmond believes that the experiments are 'systematically forfeiting a godsent chance to see the world through an ape's eyes'.

We have not done very well so far at seeing their point of view or living alongside them. We have invaded their territory and shoved most of them out of the way. While we decide what we think of them and what name, rank and serial number to give them, one title we have certainly bestowed on them is that of Endangered Species. Our closest relatives are taking their places beside many other animals with all-too familiar tales to tell about the relentless demolition of their forest homes. Chimpanzees are also hunted for meat and captured alive for zoos, the pet market and scientific research.

Chimpanzees used to live throughout Africa's tropical belt that spanned 2 million square miles, from southern Senegal through central Africa to western Tanzania. It is an area almost the size of the US and includes 25 countries. Chimpanzees are now extinct in 4 of them, and are down to such low numbers (100-200) in 5 more that their disappearance is inevitable. Another 5 countries contain small, scattered populations of a few hundred. Only 10 countries still have 1,000 or more. No one knows exactly how many wild chimpanzees there are in the world because some regions where they are known to live – such as northern Zaire – have never been properly surveyed. There are estimated to be about 200,000 left in Africa. Only fifty years ago there were probably several million.

It is not just the stark numbers that cause such concern, it is the way they are squeezed into the remaining pockets of forest. Populations that

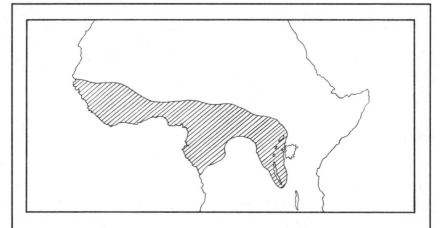

Estimated chimpanzee range 10,000 years ago...

It used to be thought that there were as many as 14 different races, or sub-species of chimpanzee, but now only 3 are recognized. Rivers keep them separate as chimps do not like water and cannot swim. The 3 subspecies are hard to tell apart though their faces are slightly different.

WESTERN CHIMPANZEE (*Pan troglodytes verus*)
Dark mask surrounding eyes, pale face, often with a white beard.

Original range: Southern Senegal eastwards to the Niger river in Central Nigeria.

Present range: Extinct in The Gambia, Guinea-Bissau, Burkina-Faso, Togo & Benin. 57% of total population lives in Guinea. Small populations in Sierra Leone, Liberia and Ivory Coast.

CENTRAL CHIMPANZEE (*Pan troglodytes troglodytes*)
Freckles on face

Original range: Nigeria east of Niger river to the Ubanghi river and south to the Zaire river.

Present range: Very low populations (a few hundred) in Nigeria, Angola, Central African Republic. Slightly higher numbers in Equatorial Guinea, Congo, Cameroon. 80% of total population in Gabon.

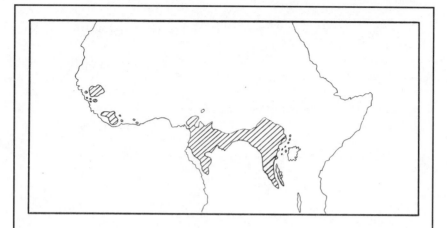

and today.

EASTERN CHIMPANZEE (*Pan troglodytes schweinfurthi*)
Bronze/coppery face. Long side burns. Not much beard.

Original range: North and east of Zaire river from Ubanghi river in eastern Zaire to the Rift Valley lakes in the east, and from southern Sudan to the southern end of Lake Tanganyika near the Zambian border.

Present range: Severely threatened (down to a few hundred) in Burundi, Rwanda, Sudan. Fragmented populations in Tanzania and Uganda. 93% of total population presumed to be in Zaire.

are broken up and scattered are unlikely to recover. Groups may be forced to live in damaged habitats, and continued disturbance disrupts their whole way of life. Even if they were left alone it would take years for them to build up their numbers again. A female does not have her first baby until she is about 12 years old; there is a 5 to 6-year interval between births, and she only bears 4 or 5 babies in her lifetime. One study found that fewer babies are born to groups in fragmented forests.

Yet fragments are all they have left. Jane Goodall writes: 'The most we can hope for in the future is to set up a series of preserves and national parks where the chimps, to some extent, can live out their natural lives. But even if we manage to do this – and unless we act soon, it will be too late – we can hope at best for a series of islands within which the chimps will be imprisoned.' Gombe is just such an island. Hills that were covered in trees thirty years ago are now bald. Similar scenes appear throughout the tropical belt. The trees end up as the doors, window frames and toilet seats demanded by people the other side of the world, and the land is mined, built on or turned into fields.

That 1-1.5% difference in our genes makes us human. Being our nearest relatives, chimpanzees are the animals we have most chance of getting to know – and we are only just starting to make their acquaintance.

MASTERS OF DECEPTION

One area currently under scrutiny is the ability to deceive. It is one thing humans are very good at. Man the Liar seems rather a dubious accolade, but you do need to be clever to carry it off. You need to be aware of your own actions and the effect they have on others, then you need to work out how to alter or conceal them, predict the outcome, and make sure all the trouble is worthwhile. Chimpanzees have already convinced most experts that their ability to be calculating and manipulative and to play one another off against each other deserves to be described as 'politics' – and some chimps, at least, are excellent sneaks and cheats, good at downright deception and at being economical with the truth. For example:

A mother, usually an avid meat-eater, shows absolutely no interest in the meat her son is relishing. She looks at everything around her but the meal as she works her way to within an arm's reach of her son who is completely taken in by her indifference and completely taken by surprise when she steals his food.

At Gombe, when bananas were kept in chimp-proof boxes, one young male worked out how to open them by unscrewing the handle of the remote control pulley. The first few times he achieved this feat he let forth loud food grunts which summoned his seniors who gobbled the lot. Then one day he assumed an air of lofty detachment from the handle he just happened to be sitting next to, and looked around at everything but the box, and the handle that he just happened to be unscrewing. Biding his time, he sat with his foot on the handle so the box stayed closed, and only when no one was around did he release the handle, run to the box, and *silently* help himself.

Another young male also decided that keeping mum was the best way to hold on to his bananas. At first he thought it was safe if he collected his share after all the big males had taken theirs, but his food grunts meant they came and relieved him of his share too. Eventually he forced himself to muffle his voice. 'We could hear faint, choking sounds in his throat,' writes Jane Goodall.

Making a Living

Top of the list of most animals' requirements are plenty of food, a safe and comfortable home and a chance to pass on their genes. For many primates, living together is the best way of fulfilling these needs. There is usually enough food to go round, and it is useful having more than one pair of eyes to find hidden fruit and shoots. There is safety in numbers too, with everyone on the lookout for predators such as snakes and big cats, and for strangers of their own species. Not all primates are strictly territorial but groups do tend to avoid each other.

For some, 'group' means a monogamous couple and their offspring, for others, it means several females and their offspring plus one or more adult males. Whatever the arrangement, the group eats, travels and rests together.

A chimpanzee may spend one whole day alone, team up with five others the next, and go off in a twosome the next. No wonder one observer described them as 'footloose'. But there is more to this than meets the eye. What looks at first haphazard is actually very sensible. Between 15 and 120 chimpanzees of both sexes and all ages live in a group, known as a unit-group or community. All members know each other but they are seldom all together because they divide into smaller groups of six or less. These 'temporary parties' last from a few hours to a few days. Their members change when parties meet, intermingle and disperse in different combinations or when someone chooses to drop out – for as well as being very sociable, chimpanzees spend a lot of time on their own. All this coming and going has been dubbed 'fusion-fission'. It is a much more complicated life than staying in a fixed group, but you can see it makes sense when you see where they live.

Africa's tropical green belt contains lush, evergreen rainforests, riverine forests, open deciduous woodlands, and mosaics of forest, wood and grassland. Most animals have to stick to one habitat, but chimps can make a living from all of these so long as there are mature fruit trees, for fruit makes up the bulk of their diet.

Rainforests offer a wide selection of fruit but often not much of each kind because there may be only a few of each species of tree and often *each tree*, let alone each species, bears fruit at a different time. Deciduous woodlands have more predictable cycles but the fruit comes all at once. The more varied seasonal gluts and scattered distribution mean that food for chimps appears at different times in different patches. Other food such as certain kinds of leaves, shoots and insects, also come in smallish portions here and there – so to make sure everyone gets enough, the community splits up and feeds in different patches, and joins up again when there is plenty to go round. A large fig tree, for example, has been known to hold over twenty feasting chimpanzees.

Each community tends to occupy one area of land but does not keep within strict boundaries so the area is described as a 'home range' rather than a territory. How much overlap there is between home ranges varies. Each home range must be large enough to support everyone, so its size depends on the number of chimps and the type of habitat. The average size is between 4-20 square miles (10-50 square km) with population densities of 0.4-2 per square mile (1-5 per square km). Some rainforests support more chimps in a smaller area, while savannas have the least food and the biggest home ranges. The hottest, driest and most sparse chimpanzee habitat ever studied is in Niokolo Koba National Park in Senegal, where there is only 3% forest cover, and 25-30 chimpanzees roam in 129 square miles (333 square km).

Chimpanzees know their neighbourhoods very well. They travel on the ground and follow a network of paths – but they do not just wander aimlessly in the hope of coming across something to eat. Chimps know where they are going. They remember from one year to the next where and when particular food appears, and each day they decide where to go to find a good meal. Many's the time baffled observers have watched them speeding up a good three minutes before the laden tree comes into sight.

Carrying this 'mental map' requires considerable brainpower, but there is more to bear in mind than where the next meal is coming from; where are they going to spend the night, for instance? Chimpanzees always sleep in trees other than the ones they feed in, so they need to make sure they do not stray too far from suitable lodgings, especially in open savanna. Water is another resource that limits where they go. Juicy fruits quench their thirst, but they also drink from streams and puddles, and need to make sure they are not left without water, especially in the dry season.

Other animals must be taken into account too. Fruit is raided by some birds and monkeys but the thieves soon scarper when they hear chimps coming – and any caught in the act are chased away, although chimpanzees usually withdraw when big baboons scream and charge at them. Lions, leopards, monitor lizards and crocodiles are the most likely animals to kill and eat chimps, and probably pick on the old, the young and the sick. Until recently hardly any attacks had been witnessed though Dr Nishida has collected conclusive evidence that chimps at Mahale have been eaten by lions. In April 1989 some lions moved into Mahale and during the next few months three young chimps disappeared. Chimp remains – hair, skin, bones and teeth – were found in lion faeces. Dr Christophe Boesch has seen or heard a number of confrontations between leopards and chimpanzees in the Tai Forest, and believes that leopards are the main cause of death to chimps there. He has seen four chimps killed and six injured by leopards. Eleven victims were rescued by fellow chimps who chased away the leopard, and in nine cases the leopard was seen off before it had a chance to attack. Elsewhere, adult chimps have been seen pelting snakes with sticks and stones, but they usually turn tail at the first sight or sound of danger.

Injury or sickness of the juveniles slows a party down and dictates where it goes, and some parties travel much further than others. Although individuals mingle freely, party membership is not entirely random. Basically, males are more gregarious than females, so all-male parties are very common, while females often stay on their own or just with their off-spring. A youngster depends on its mother for at least 5 years, and even after reaching adulthood at about 13 years old, it spends a lot of time with her. A female has a baby every 5 or 6 years, so she has several children of different ages in tow.

The demands of pregnancy and lactation make it important for a female to keep healthy, and in order to get enough food for herself and her family without having to trek miles to find it, each female tends to stay in her own mini home range or 'core area'. Core areas are distributed throughout the home range and overlap each other. Two or more families form temporary parties when they cross paths, and stay together longer if there is enough food for everyone.

Not tied down by children, the males seem free to roam the whole of the home range. At Gombe, a female travels about 1 and 2 miles (2-3 km) per day, while a male covers about 2½-4 miles (4-6 km). One male once walked over 6 miles (10 km), but took the next day off, only managing about 1 mile (2 km)! But the males are not the free agents they might appear. In fact, where they go and what they do is largely in the hands of the females: while a female's best chance of passing on her genes is to take good care of herself and her children, a male's is to mate with as many different partners as possible, so the males criss-cross the range and make sure they meet all the females regularly. Their movements are also designed to protect the females and young and to patrol their territory.

Although no two days are exactly the same, chimpanzees do follow a rough routine, dividing their time between eating, travelling, resting and being with each other.

Chimpanzees wake at daybreak; usually their first activity is to set off to look for breakfast. Some primates snack all day, but chimps prefer to fill themselves up and then take a breather. The first meal of the day can last several hours. Then they while away the hottest hours, sitting around and snoozing, and get going again in the afternoon. After another

GOOD DOCTORS

It seems that chimpanzees might be good doctors as well as good botanists. They not only eat soil to aid digestion of certain plants, but also eat certain plants that act as antibiotics. Generally they eat fruit in the mornings and leaves later on, but every so often the chimps at several different sites have been seen to start the day with the leaves of an herbaceous plant called Aspilia. Some days they walk right past these plants without a second glance, but on others they make special journeys to find them. And whereas most leaves are picked and chewed without much ceremony, these are carefully selected: each leaf is scrutinized and then held between the lips. Some are rejected and left on the plant. Others are bitten off and rolled around in the mouth, then swallowed without chewing, despite being dry and bristly.

Some African people use these same leaves as medicine, mostly for stomach troubles, including worms. Chimps eat them more during the rainy season when they are more susceptible to worms, and laboratory tests have shown that they do contain antibiotic properties that can kill some parasites.

Chimp picking medicinal plants.

Group of males hunting.

expedition and another big meal they are ready to take to the trees for the night.

Sleeping trees must have plenty of foliage and bendable branches, for each chimp makes a fresh nest almost every night (and sometimes day nests are made in trees and on the ground, for the midday siesta). Leafy branches are folded down and held in place with the feet while another layer is piled on top and tucked round the sides. The result is a springy platform that provides comfort and shelter. Snakes and leopards climb trees, of course, but the rustling and swaying of branches might act as early warnings. Some chimpanzees break off branches overhanging their nests – perhaps to prevent predators dropping in. They are watchful as they settle down, and several parties often cluster together with up to ten nests per tree.

Early risers as a rule, chimpanzees are reluctant to get up when it is raining. Their coats are not very waterproof and they do not like getting wet. They nest earlier in wet weather, and make more day nests. If heavy rain catches them on the move most of them duck under trees or hunch themselves up and sit it out, but adult males sometimes do the opposite: wind, rain and rushing water can all trigger a male to display. He starts by rocking gently, then stamps his feet and waves his arms, and finally charges round throwing branches. The effect is made even more impressive by the fact that all his hair stands on end. No one knows the reason for this 'rain dance'; it is as if he is raging at the elements.

Chimpanzees spend more time feeding in trees during the rains – partly to avoid the wet ground, partly because less food grows on the ground plants during the rainy season. It is easy to sit in a tree and gorge ripe fruit, but chimpanzees sometimes have to work hard for their living, walking from place to place and making do with flowers, stems and bark.

Although fruit is their staple, it lacks protein and certain nutrients, so to get a good balanced diet they must eat fresh leaves and shoots, and some meat – and will abandon a banquet of figs in favour of this other food. One study found that a chimpanzee's annual intake averages 68% fruit, 28% leaves and other plant matter, 4% meat. Chimps eat 150-200 species of plant and over 20 different insects, including ants, caterpillars, crickets and beetles. One day's menu can include up to 20 different items. And it is not just variety they are after: they will seek something for its nutritional value and their particular needs. After watching them making careful selections of plants and parts of plants, Dr Richard Wrangham described the Gombe chimps as 'good botanists'.

Having chosen what they want, individuals vary in their eating habits. Fussiness partly depends on how hungry they are: a hungry chimp wolfs

Tool use: cracking nuts with a stone.

down skin, pips and all. Some food takes time and trouble to prepare, collect or extract. Tiny berries are picked gingerly from prickly bushes, protein-rich seedpods are split with the teeth, prised apart with thumb and forefinger, and the seeds picked out with the lips. Teeth also break through thick-skinned fruits, and hands and feet are used to rip and tear. Good though it is, this equipment is not always up to the job. Hard-shelled fruits and nuts are smashed against rocks or tree-trunks. But the chimpanzees of West Africa use more sophisticated techniques. They use nutcrackers.

More precisely, they use hammers and anvils to crack nuts. When Dr Boesch first set out to track down chimpanzees in the Tai Forest, it was the sounds of their hammering that gave them away. Since then he has made a thorough study of nut-cracking. Hammers are made from logs or stones, anvils from stones or exposed tree roots. Five species of nut are cracked, and some are harder than others, so logs are broken to the appropriate size, and stones are carefully selected. Hammers range from 280 g (10 oz) to 20 kg (45 lb). Stones are rare in the forest and the chimps are not only good at finding them but keeping them: they remember where they left a particular tool and go back to it time and time again – and even take it from one nut-cracking site to another.

Sites are at the foot of the nut trees, and abandoned ones are strewn with empty shells and tools. The chimps sometimes take their hammers up a tree, pick the nuts and crack them up there, but usually they gather hand- and mouthfuls from the ground and take them to favourite anvils (worn smooth into a dip). At the height of the nutting season they spend 2-3 hours a day nutcracking, each getting about 270 nuts.

It is a highly skilled job. A nut must be hit at just the right angle and hard enough to crack its shell but not so hard as to pulverize the kernel. Dr Boesch writes: 'Time and time again, we have been impressed to see a chimpanzee raise a 20lb stone above its head, strike a nut with ten or more powerful blows, and then, using the same hammer, switch to delicate little taps from a height of only four inches.' And then, to prise out the bits that always stay inside, they use sticks held in the mouth or hand.

Sticks are essential for other delicate extractions such as fishing for termites. Baboons and birds also regard termites as tasty snacks, snapping them up as they emerge from the mound on their nuptial flights. Chimpanzees, however, get a head-start by winkling them out of their mounds. Jane Goodall first saw chimps collecting termites by inserting probes into their mounds and broke the news that man was not the only tool-maker – and as if to rub it in, chimps are better at this particular operation than man. Or one man, anyway. When Geza Teleki was at Gombe he studied the chimps termite-fishing for months. But when he had a go himself he failed miserably. The first thing you have to do is find one of the hundred or so mound openings. They are invisible because the workers seal them. The chimps have little trouble, and quickly scratch them open, but Teleki could not find them. Next you have to choose a stick or stem pliable enough to go down the crooked corridors of the mound, but not too flimsy, and then you have to prepare it by breaking it to the right length and pulling off any leaves. Once you have inserted your probe you have to wriggle gently to bait the prey, then wait, then withdraw quickly and smoothly so as not to tear the grip of the termites that have fastened on. Then you have to eat them before they bite you. Teleki just could not get the hang of it.

Eating angry insects calls for ingenuity. Driver ants, for example, live underground and do not like to be disturbed. Chimpanzees disturb them with caution and keep out of biting range by inserting their probes and standing back while the ants crawl up the sticks. Some 'ant-dip' by dangling from a branch above the nests – but this carries the risk of the branch breaking.

Without their dip-sticks, probes and hammers, chimpanzees could not eat many of the foods they rely on. There are other famous tool-users, such as the sea otter that lies on its back, puts a mollusc on its chest and smashes it with a stone, or the finch that digs out insects by holding a stick in its beak – but they use one tool to do one job and repeat the same process over and over again. Chimpanzees are less automatic. They choose a tool carefully, prepare it, then try it out and adjust it. And they plan ahead, collecting several probes on their way to a mound, or fetching a favourite hammer. Not many animals can keep their attention on anything for long, but chimps really concentrate on the job in hand. Termite-fishing or nut-cracking can easily last an hour or more without a break.

They also put time and effort into getting meat. They steal it from baboons, and scavenge it, but mostly they hunt. Prey includes mice, rats and nestlings, a variety of monkeys, fawns, and hefty half-grown bushpigs weighing 20 kg (44 lb). Animals are grabbed in passing (described by Russell Tuttle as 'opportunistic lunges'), caught in high-speed chases, or cornered in well co-ordinated hunts lasting up to two hours.

In general, females do more opportunistic lunging and males get together to stalk and chase. One of the first proper hunts ever seen involved two young male chimpanzees and an adult red colobus monkey. One chimp got the monkey's attention by climbing the tree next to it while the other climbed the monkey's tree, crept up on it and grabbed it. Since then some remarkable co-operation between five or more males has been witnessed, especially among the monkey-hunters of the Tai Forest. Christophe Boesch describes how a group of males may spot monkeys in the distance and go after them, or just scan the trees until they find some. They climb together, then split up, some stalking their victims, others placing themselves in strategic positions to block escape routes. Then the stalkers close in and herd the monkeys into the arms of the ones waiting in ambush.

At which point the silent, controlled hunters turn into frenzied maniacs, shrieking and displaying, and crowding round whoever has the prize. Unlike carnivores, chimps are inefficient killers. They may break the neck with a bite, but otherwise batter and rip the victim to death, and eat it alive. They chew meat very thoroughly, and add leaves or other fibre to help it down. Hunters get the lion's share, but community members who beg for pieces with outstretched hands are often rewarded. Sometimes a chimpanzee will hand out morsels without even being asked – something unheard of in the animal world apart from parents feeding offspring.

To be off on their own one day and working closely together the next is the most sensible way for chimpanzees to make a living. The fusion-fission system gives them maximum flexibility to suit their ever-changing ecological scene. But it also makes for an ever-changing social scene, and for this to work the individuals involved must have a really good understanding of each other.

TOOL BOX

Food utensils are the most common, but chimpanzees use tools for many other purposes. Leaves are used as cloths to wipe dirt off the body, or can be chewed up to make sponges that are soaked in water and sucked to get a drink. Big sticks and stones are used as clubs and missiles in fights with other chimps and to scare off predators. Smaller sticks are used to investigate anything unfamiliar without actually touching it (including newborn babies!) and even as toothpicks.

Chimpanzees are always examining the things around them. They sniff, handle and pull apart bits of bark, leaves or roots, and every so often they hit on a new way of using an object to solve a problem. Jane Goodall describes how an adult male, Mike, wanted to take a banana from her hand but was too wary to touch her. In frustration he threatened her by shaking a clump of grass, one blade of which touched the banana. When he noticed this he let go of the grass and picked up a thin plant then discarded it for a stick which he used to knock the banana from her hand. When offered a second banana he simply used his new tool.

OPPOSITE: *angry male.*

• 4 •
Friends and Relations

Chimpanzee reunions can be pandemonium. When parties meet after several days' separation everyone rushes about screaming, and in all the noise and excitement it is hard to tell whether they are glad to see each other or not. It is also hard to know where to look: everything happens at top speed, and it seems like one big scrum. But a chimpanzee can break the chaos into a series of clear messages about everyone's moods and intentions.

Chimpanzees communicate with sounds, facial expressions, postures and gestures. If you saw a chimp glare at you, stand upright, scream and charge towards you with all his hair on end, you might make an intelligent guess that he was angry. One sitting hunched up, pouting and whimpering is quite clearly upset. From fear and anger to joy and contentment, chimpanzees experience a whole range of emotions – and wearing their hearts on their sleeves means that it does not take a human being long to learn how they are feeling.

But they also express much more subtle emotions, and constantly swap information about themselves and their surroundings by means of a grunt, a glance or a slight change in posture. Even when they are sitting around doing nothing they are busy sending and receiving signals. Eye contact, for example, is an important way of opening the communication channel: one chimpanzee – with aggressive or friendly intentions – tries to catch the other's eye. The other may deliberately look back, or deliberately look away. A crash course in chimpanzee communication would teach you everything to look out for, but really to know what is going on you need to know all the individuals involved, their history and relationships. At least, that is what the chimps know – for in a group each individual is affected not just by his or her own relationships, but by everyone else's too.

What with strong streaks of independence and a highly sociable nature, chimpanzees sometimes seem to be pulled in two directions, and when they are together they need a few rules and regulations to help them organize themselves. Having a hierarchy keeps them in order, but they do not just automatically slot into a fixed place; the hierarchy is not a rigid framework imposed on them, but a loose arrangement of jostling individuals. And since rank is not inherited, they are always checking where they stand with each other, and either re-affirming the relationship or challenging it. Much of the action is hardly noticeable, with one chimpanzee simply moving out of the other's way at a feeding site or sleeping tree. But when they have not seen each other for a while they need to find out if anything has changed – hence the reunion hullabaloo.

In a typical encounter A strides about with shoulders hunched, mouth

Typical greeting between dominant (on left) and subordinate.

clenched and all his hair standing on end. Sleek-haired B darts backwards and forwards in front of A, with one arm outstretched towards him. Without taking his eyes off A he pant-grunts at him – a quick series of breathy grunts that veer into screams – then turns his back and crouches in front of A, looking at him over his shoulder with his mouth split in a wide grimace. A walks past B, soft-barks, and puts a hand on his back, which quietens him.

Everything about them tells you (and them) who is dominant. A oozes confidence and makes himself look bigger than he really is, B bows and scrapes and has vulnerability written all over him. Submission is most commonly expressed by crouching and backing towards the superior (known as 'presenting') and by pant-grunting which is only ever directed up the hierarchy. Soft-barking is only ever directed downwards, and arm-waving, bipedal swaggering and hunched shoulders are all expressions of dominance, made even more impressive by bristling hair. These gestures also feature in full-scale charging displays when a chimp thunders around at speed, screaming and throwing branches. Displays are the most impressive demonstrations of authority, more often aimed at fellow chimpanzees than at rain and wind.

Subordinates may just get out of the way, but they often want to be noticed: to pay their respects and receive reassurance. The pat on the back tells B what he wants to hear, and afterwards he is likely to sit near A. The hierarchy keeps the chimpanzees together: once they have acknowledged where they stand with each other they can share the same space peacefully.

Hugging.

Only when B stands up to A does the trouble start. Dominance disputes range from screaming matches to displays, and it may take several rounds over a number of weeks before anyone backs down. Fights do sometimes break out, and adult chimps are far stronger than human beings. They slap, grab and stamp on each other, and their long canine teeth can inflict severe injuries. Thanks to their displays, they often manage to vent their anger without causing much physical damage. Expressions of friendship, on the other hand, rely on physical contact.

Amidst all the charging about at reunions there is a lot of hugging, kissing, patting on the back – and when the excitement dies down, the chimpanzees often embark on a grooming session. Running their fingers through each others' hair removes dirt, dead skin and parasites, but its value goes way beyond hygiene. Grooming is the single most important social activity and takes up a lot of each day's rest periods. Via grooming they patch up disagreements, reassure each other, affirm friendships and simply relax together.

A chimpanzee asks to be groomed by looking at someone and ostentatiously scratching an arm or leg, or offering up a part of the body. Pairs sit or loll together, each grooming the other or taking it in turns. Tempo varies from languid ruffling with a forefinger to brisk parting with both hands, and is often accompanied by smacking the lips.

As the session progresses, those on the receiving end get more and more relaxed and even drift off to sleep. They trust each other enough to allow their arms, legs or heads to be moved out of the way, and to be groomed round delicate areas such as the eyes or groin. Most sessions last about an hour. Longer ones usually involve more chimps – up to ten get together in one grooming group.

Adult chimpanzees have a highly developed sense of reciprocity, and although high rankers are often on the receiving end of grooming, they give as well as take ('you scratch my back . . . '). Grooming behaviour also shows how socially sophisticated – or manipulative – chimps are, for they will offer to groom someone in order to get what they want. For example, males may groom females when they want to mate with them, and mothers of young babies are groomed by nosey chimps who want to get a better look at the newcomer.

Individual personality plays a large part in how well any two chimpanzees get on, but adults always outrank youngsters, and adult males outrank adult females.

A female wants to be left in peace to bring up her family, and her main aim is to establish herself in a good core area with plenty of food. Once she has achieved this, there is not much point in trying to outrank other established females. But she does cause friction on her way.

It begins when she first leaves home. Promiscuous societies need to guard against in-breeding so young adults of one or other sex (or both) leave the group they were born in. With some of the biggest chimpanzee communities no one leaves, but often the late adolescent and young adult females of 10-12 years old move out and join neighbouring groups.

Before taking the plunge the female makes short visits to get to know her new home and its inhabitants. Resident females give her a frosty reception, chasing and biting her. They see her as a threat because there is only so much land to go round, and if she sets up home in their range she will be invading their space and depriving them of food. The males see her in quite a different light. To them she is welcome as a potential sexual partner, and a female usually goes visiting when she is sexually receptive. For two to three years a young female has long periods of receptivity but her body is not ready to conceive this attractiveness seems to be her entry ticket, and she travels with the males who keep resident females off her back.

It takes months or even years fully to transfer, and some try more than one group, or go back home after a spell away. As the new girl she has the lowest status of all the adult females. Her core area is unlikely to contain the best fruit trees, and is often on the edge of the range. Her status improves with age, especially if she has a lot of children to support her in disputes. The highest ranking females are the eldest with the biggest families.

Most females rub along together and are so good at minding their own business that they can share a tree without acknowledging each others' existence. But they do strike up special friendships, sticking up for each other in arguments, and getting together whenever they can afford to – at times and in places where food is abundant. Relationships between adult females are often low-key affairs, and it is easy to overlook just how much does go on between them because the males are so much more dramatic.

Males are certainly more intense. They argue more, co-operate more, groom each other more, spend much more time together, and make much more noise. And their part of the hierarchy is much more complicated because they are caught between two urges: to compete and to co-operate.

Males have an obvious drive to reach high rank. A male enters the status game as a young adult of about 10-12 years old and climbs steadily until he reaches the peak at about 20-25 years old. For the next 10 years he is known as a prime male, and even if he does not actually become the dominant male, he stays in the upper echelons until his 30s. After that he starts slipping, but he is regarded as a kind of elder statesman: the higher ranking younger males often give way to him, share meat with him, and groom him. He enjoys these privileges until he dies of old age at about 50 years old.

Males display at each other and sometimes fight, but they are not bullies or despots. Top rankers owe their positions more to their personalities than to their muscles. The dominant male is often not the strongest or the most pushy, but the one best at commanding respect and keeping the community together. Even when parties are most scattered they listen out for his voice and wend their way towards him. Though they compete, the males also need each other. They are like acrobats forming a pyramid: standing alongside each other and holding on to each other to support the next tier balancing on their shoulders. And it is wobbly at the top.

High rankers have to work hard to persuade their supporters to stay loyal, and they forge close friendships with one or two other adult males on whom they rely. They travel and feed together, groom each other and back each other in arguments. These coalitions last months or years, and the strongest and longest-lasting are between brothers. They also can be the basis of power takeovers. On his own B would not dream of standing up to A but together B and C can challenge him. This is especially important if B is on his way up, and can make all the difference to whether he overtakes A or not. It has even been known for two in a power struggle to engineer it so they only meet when they are with their allies.

The dominant male often has an entourage of three high rankers which does not include the sub-dominant male. He has his own supporters. The relationship between the two top rankers is tense at times of change, especially at reunions when they and their allies test whether anything has changed between them. So aware of the situation is he that the alpha tries to split up the beta's allies, charging at them while they sit together. At the same time he puts a lot of effort into keeping his own followers loyal, sitting near them and grooming them *more* than they groom him. One clever strategy used by some dominant males is to ally with elder statesmen who can usually threaten up-and-coming young males without retaliation, thus keeping the dominant male's rivals in check.

Not all males want to get right to the top. Some prefer to stay close to the top ranker rather than risk challenging him. An ally is in a powerful position as he can play the two top dogs (chimps) off against each other. He can change his allegiance, and support whoever is most dominant, or he can to and fro between alpha and beta, since both need his support. One wily elder at Gombe shifted between two dominant males six times in three months.

All this rivalry is balanced by friendly contact, and the members of different cliques are quite amicable. One study found that males groom each other four times as frequently as females do, kiss each other twenty times as much, and that 80% of all hugs were between males.

Such a busy social life takes up a lot of time. When many males are together they spend less time eating than when they are alone or with one companion. Left to their own devices much more, and often having more than one mouth to feed, females at Gombe spend up to twice as much time as males collecting insects, and the Tai Forest females are consistently better nut-crackers and work at it for longer. Females seem to have more patience and better powers of concentration, sticking at the job in hand even when they do not get much reward for their labours.

The privileges of high office are not always obvious, and do not even guarantee a full belly: when food is scarce and the chimpanzees split up, high rankers are not necessarily better off. And when it comes to meat – the most highly prized food of all – the alpha has to do his share of begging, for possessors of meat give most readily to their allies. But the most important prize is a female – and the most dominant males have the most opportunities to mate.

Adult females have an oestrous cycle of 27-39 days and are receptive

Female with swelling.

for 2-3 weeks every 4-6 weeks. Receptivity is signalled by a swelling in the perineal skin (which only lasts 10 days). Contrasting with their black hair, this pink cushion can be spotted some distance away, and has been described by Goodall as a 'visual aid'.

The male usually makes the first move by trying to attract a female's attention with a courtship display. Described by the experts as 'non-vocal attention-getting' or the 'male-invite posture', he sits with his legs open to reveal his erection, and gazes at her. Added extras include rocking, shaking branches, stamping a foot, or hitting the back of the hand on the ground. Mahale males have some techniques of their own. One is to make a rough nest in a tree or on the ground and invite her into it by sitting in it and stamping a foot. Another (also seen at Bossou in Guinea) seems to be a combination of frustration and attention-seeking, and is known as the leaf-clipping display. A male picks a stiff leaf, holds the stalk in his hand, and noisily bites off small pieces and spits them out. When he is left with nothing but the stalk and the main vein he drops it and starts on another. Most leaf-clippers are young males trying to catch a female's eye without the older males noticing. Some young males caught leaf-clipping by their seniors then pretend they are just making a fishing tool or eating the leaf!

Although a male sometimes stands upright and swaggers over to her, he cannot force a female to mate with him. She rejects his advances by ignoring him or moving away. She accepts them by crouching in front of him with her back towards him, in a posture known as crouch-presenting. Females also initiate sex by backing into a male. The commonest mating position is for the male to squat behind the crouching female, and they often do not touch each other during copulation. While he thrusts, the male pants and the female squeals, both building up to a crescendo.

The first week of a female's swelling is the most promiscuous time, when males mate freely in front of each other. Although they do not have consistent partners, chimpanzees do show preferences, and older, experienced females are often the most popular. Young ones can be nervous, and leap away before the male has finished. Females crouch-present most to high ranking males and often stay near them. When they are in oestrous, females are groomed much more by the males, and are even allowed to take food from them.

Maximum swelling means the female is nearing ovulation, and it is then that the males pull rank, chasing and threatening subordinate rivals. The dominant often fends off *everyone* else, ensuring that he is most likely to make the female pregnant.

By following her around and grooming her, the dominant male does his best to keep tabs on the female but he needs her co-operation if he is to keep her to himself, and his rivals do their best to lure her away from him: one of the other males makes his intentions clear by staying as close to her as he dares, and gazing at her. Then he seizes his chance when everyone gets up to go, and walks off in the other direction, glancing over his shoulder to make sure she is following him. If she is not, he may sit and stare at her or shake branches at her, but he does not want to draw too much attention to himself. Some cannot help it, and break into a display, which usually blows their chances as it brings the wrath of their rivals down on them. A female is most likely to go with a male who has groomed her and shared food with her during the promiscuous phase. Some have been known to go along and then change their minds, answering the calls of the other males who come and 'rescue' them and beat up their beaux.

Couples who do sneak off form what is known as a consortship, and stay together a few days or even weeks. They often travel to the most isolated parts of the range – which is why they are also known as being 'on safari'. Most couples are relaxed together and indulge in long grooming sessions. If a male can get away with it, a consortship is his best chance because a female is more likely to become pregnant during a consortship than during the promiscuous phase. But there is a price to pay. As Russell

Grooming.

Tuttle puts it: 'Absence makes the heart grow harder.' If he is not there to join in the status games he may come back to find he has lost his place in the hierarchy. Being able to stay in the thick of things *and* keep a female to himself means the alpha can have his cake and eat it.

For all their sexual rivalry, many of the males in one community are related, and it is better to lose out to the devil you know – who might just be your nephew or your cousin – than to a complete stranger. Males put such a lot into keeping on friendly terms with each other so they can present a united front to outsiders and keep them away from 'their' females.

Chimpanzees are not strictly territorial in the sense of having exclusive land rights, although how much overlap there is between ranges varies in different places: ravines and rivers make natural barriers, and some communities recognize what Dr Nishida calls 'invisible barriers'. Few communities stay in exactly the same piece of land year in year out, for ranges expand and shrink according to the changing size of the community.

Male mistrust of strangers can mean anything from keeping out of their way to picking a fight. When foraging in the outskirts of their neighbourhood, the males are on edge: any sudden noise gives them a jump, and they sniff recently trodden leaves or discarded tools to pick up clues about their neighbours. Sometimes they announce themselves with loud hooting and shrieking. Strangers stop whatever they are doing and listen, then shout back. These vocal contests may be enough to persuade each party to keep its distance – or they may force a confrontation.

Many face-to-face encounters are nothing but displays. If one team is larger than the other, the smaller one usually backs down and goes away. More evenly matched teams may also resolve matters by retreating; they seldom risk fighting unless they clearly have an unfair advantage. Most attacks are on lone individuals.

With no fixed travelling patterns, neighbours are bound to bump into each other, but not all meetings are by chance. Sometimes the chimpanzees go looking for trouble. Every four days or so, parties of males go to the perimeter of their range. On these 'boundary patrols' they behave quite differently from when they just happen to be there to forage. Patrollers stick close together, creep along in silence, and even avoid treading on dry leaves.

If a glimpse of their neighbours is all the spies are after, they slink away once they have got it (and burst into voice once safely back home). If a show of strength is their intention, they suddenly reveal themselves and put on a display, watch their enemies display in reply, and leave noisily. But they may be out to take over land and recruit new females. Some invasions have led to very bloody battles, and both Goodall and Nishida have witnessed whole communities systematically wiped out.

As with the chance encounters, the attackers minimize their losses by picking on lone males or small parties, making a protracted series of onslaughts. As when going hunting, they divide their labour in order to get the job done, some holding down the victim while others stamp and bite. It is estimated that a quarter of all adult males that die at Gombe are murdered. Here males turned on neighbours who had been fellow members of their community. A few years previously their group had split into two and taken up residence in different parts of what had been their common range. Then the males of the larger community in the north (known as the Kasakela community) attacked males of the Kahama community in the south. Over the next seven years all the Kahama males disappeared. Some adult females were also killed, but not their adolescent daughters. These and any other surviving females and their offspring joined the victors who spread across the Kahama range.

Victory for the Kasakela clan turned out to be short-lived. Within a year a powerful community advanced from the south, pushing them out of the Kahama range and further north than they had been originally. There they were stopped by another group advancing south. They were sandwiched, their males began to vanish, presumed killed, and their community shrank. But a few years later it was expanding again, thanks to some up-and-coming young males who pushed back the frontiers with bluff and noise, not bloodshed.

What flicks the switch and turns them from innocuous show-offs into brutal killers no one really knows. Some blame the observers at Gombe and Mahale for allowing numbers and tensions to rise by putting out food for the chimpanzees in order to get closer to them. Big, regular, good quality meals are not what chimps are used to, but, true to their fusion-fission system, large numbers converged and even hung around waiting for the food – which inevitably meant frayed tempers. To stop this happening, rationing was brought in at Gombe, and soon afterwards the Kasakela/Kahama split occurred.

Not until more long-term studies have been carried out will we know how big a part provisioning played. Jane Goodall describes how two females may have indirectly contributed to the violence at Gombe: over a period of four years, Passion killed and ate many newborn infants in the community. Again, no one knows why. Perhaps it was to keep the mothers away from a core area that Pom, Passion's daughter, was trying to take over next door to her mother. The repercussions were immense: the infant-less females came into oestrous and mated, which meant a baby boom, which in turn meant no receptive females for the next few years. Did the males at Gombe attack their neighbours with the *intention* of recruiting females?

And whatever it was that pushed them, did they really need to go to such extremes? 'Hell knows no fury like a riled chimpanzee,' writes Russell Tuttle, and this bloodshed certainly shows how vital their usual constraints are; how their displays are safe ways of expressing anger, and how – amongst themselves – aggression goes with appeasement and reconciliation. Faced with enemies only the aggressive side comes out; without any checks it can boil over.

Dr Goodall discusses whether the Gombe males actually intended to kill. Several times she witnessed how they did not leave their victims alone until they were sure he (or she) was immobilized. She concludes: 'If they had had firearms and had been taught to use them, I suspect they would have used them to kill.' The warriors certainly shattered the image of the gentle, peace-loving chimpanzee, and provoked many a debate on the nature of our closest relative.

READ MY LIPS

Pulling faces at each other is an extremely important part of chimpanzee communication. Chimps have naked faces, big eyes and flexible lips, and they have a set of expressions whose meanings everyone understands. For example, a wide open mouth with the top lip covering the top teeth is a 'play face', used during a game or to invite someone to start a game. In stark contrast is the open mouth 'fear grin' when the top and bottom lips are pulled back to show as much of the teeth and gums as possible. Then again, the loosely closed mouth of a relaxed chimp is very different from the clenched lips of an angry one.

Facial expressions and sounds often go together. Chimpanzees have a basic repertoire of 13 barks, grunts and screams, but each has its own variations according to the situation and the intensity of the emotion.

Pant hooting.

Pant-hoots are the loudest and most distinctive sounds, often used to keep in contact when parties are out of sight in the forest. Big males especially announce where they are by slapping their hands and feet on the buttress roots of trees, and uttering the spine-tingling hoots by emptying their lungs rapidly with a loud `oooo' sound and immediately inhaling noisily. The series of hoots ends in a deep roar or scream, and the noise carries at least half a mile (1 km). Parties frequently stop and listen, then answer. Chimps recognize each others' voices and carry on long-range conversations. When two parties nest within earshot they sometimes break into voice in the middle of the night. The pant-hoots also serve as dinner-gongs: a party calls when it finds a lot of food, and the others soon come running.

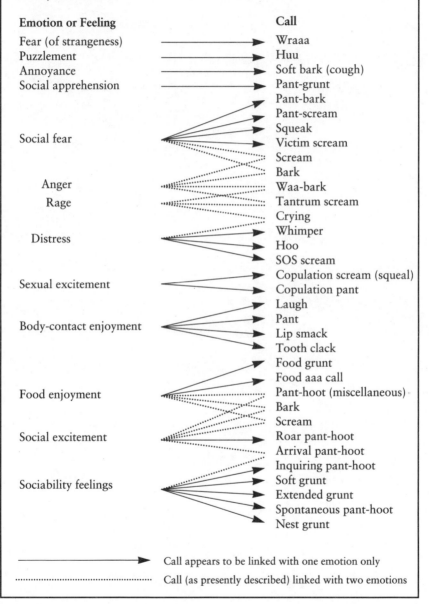

Chimpanzee calls and the emotions or feelings with which they are most closely associated.

Emotion or Feeling	Call
Fear (of strangeness)	Wraaa
Puzzlement	Huu
Annoyance	Soft bark (cough)
Social apprehension	Pant-grunt
	Pant-bark
	Pant-scream
	Squeak
Social fear	Victim scream
	Scream
	Bark
Anger	Waa-bark
Rage	Tantrum scream
	Crying
Distress	Whimper
	Hoo
	SOS scream
Sexual excitement	Copulation scream (squeal)
	Copulation pant
	Laugh
Body-contact enjoyment	Pant
	Lip smack
	Tooth clack
	Food grunt
	Food aaa call
Food enjoyment	Pant-hoot (miscellaneous)
	Bark
	Scream
Social excitement	Roar pant-hoot
	Arrival pant-hoot
	Inquiring pant-hoot
Sociability feelings	Soft grunt
	Extended grunt
	Spontaneous pant-hoot
	Nest grunt

→ Call appears to be linked with one emotion only

···· Call (as presently described) linked with two emotions

Growing up

Swaying from side to side and stamping her feet, a young chimpanzee swaggers over to her elder brother and dives into his lap. He responds to this invitation to a game by opening his mouth wide in the typical chimp 'play-face' and biting the back of her neck. Soon the pair collapse in a heap of flailing arms and legs from which rise huffy chuckles – the unmistakable sound of chimpanzee laughter

A discreet distance away their mother dozes, both ears and half an eye open to make sure her son does not get too rough with his baby sister. When the game peters out the youngster clambers back on to her mother to suckle and sleep. At two years old she is never out of sight of this source of comfort and security. Their bond is strongest during an infant's first five or six years, before the next baby appears on the scene.

A large chunk of a chimpanzee's life is spent growing up. Not until it is about 13 years old does it enter adulthood. These early years may look like an endless round of fun and games, but a chimpanzee has a lot to learn if it is to become a well balanced adult. It must not only find out about the day-to-day necessities of life such as what to eat and where to find it, but must also get to know everyone, and pick up all the dos and don'ts of social behaviour.

It all happens very gradually. The world of a newborn chimpanzee does not reach beyond its mother's hairy chest. There, the tiny (0.9-1.8 kg/2-4 lb) baby finds all it needs: food, warmth, protection. For the first days of its life its body is cupped in its mother's hand as she holds it against her. Then it manages to take four fistfuls of her hair and hang on by itself, even when it is fast asleep.

The new arrival is a source of great interest to the others, especially its nearest brother or sister, who peers at it, and tries to groom it. The mother is reluctant to let them too close, diverting their attention by playing with them or grooming them. One persistent juvenile female got her own back by grooming her mother while she investigated the baby with her foot!

At first the baby is little more than an extension of its mother: through her it feels the hugs of friendship, the slaps of anger, the pats of reassurance. Chimpanzees are born with the basic set of sounds, facial expressions, gestures and postures, but they need to experience them in action to learn how to use them appropriately. By about three months old the infant starts trying things out for itself, softly echoing its mother's pant-grunts as she greets the alpha male, or reaching out a skinny arm as she hugs another chimp. Soon after this the mother allows her older children and close friends to greet and groom the baby, and even gently nuzzle and prod it in play.

Most chimpanzees – male and female, relations and non-relations – want to greet and touch young babies, but always ask permission first by grooming the mothers or sitting beside them. Adult males are often the most polite, and are gentle when the baby starts scrambling off its mother at a few months old. Jane Goodall writes: 'It seems that a male cannot resist reaching out to draw an infant into a close embrace, to pat him, or to initiate gentle play.'

Mothers with babies under a year old tend to stay close to adult males or link up with other mothers in 'nursery groups'. Lone adults are more vulnerable to attack from neighbouring chimpanzees, and babies have been killed during fights. Some male infants have been deliberately attacked and killed by adult males from next-door communities – presumably to cut down future competition for land and females. Harder to understand is infanticide within the community. The only known female baby-killers are the mother and daughter who systematically killed and ate all newborn infants at Gombe, but sometimes adult males get together and kill babies – which seems very odd because one of them is probably the father. Dr Nishida has suggested that they thought their victims were fathered by strangers as some females do not transfer permanently, but to and fro between communities.

One mother once accidentally killed her own baby, because she did not know how to look after it. Females practise with younger siblings and other babies in their group, but sometimes their first child ends up as the guinea pig. The worst mother ever seen at Gombe threw her newborn baby on to her back, picked him up by one leg, and let his head bump along the ground when he was on her chest. Finally she squashed him by tucking him between her thigh and groin (the 'trouser pocket' where food is sometimes carried). The baby was dead within a week. Nor was she much better with the next one, but it survived, and the third one was cherished by its mother.

The personality of its mother has an enormous influence on the baby, and her behaviour with other chimpanzees is bound to rub off on it. Some mothers are relaxed and playful, others are more nervous and protective. A female's age and rank play a large part in how confident she is: an older, high-ranking female is more secure and assertive with other community members and is probably an experienced mother – which also means the baby is surrounded by a circle of babysitters, playmates and role models. The first-born of a young, low-status female spends much more time alone with its mother and witnesses her timidity with other chimps. Although rank is not inherited, early experience shapes the youngster's behaviour. When they come to making their own way in the

Baby taking food from its mother's mouth.

world, secure and confident offspring of high-status females have a head start.

The baby's first adventure into the outside world consists of an expedition onto its mother's back. It takes a few attempts to reach the right place and not slip off, but 5-6-month-olds ride with aplomb. There is a much better view from up there and, though the baby starts to walk at about 6 months, it spends most of its first two years being carried, and cadges lifts for several years after that. Being carried sounds passive, but the alert young chimp is actually very busy: looking, touching, beginning to make sense of the world around it. Chimpanzees are born with an endless supply of curiosity.

One source of fascination is food. Chimps depend on their mother's milk for 2 to 3 years, and comfort-feed for another year or more, but they start to eat solid food at about 4 months old. Babies chew on the end of whatever their mothers are eating, take food out of their mouths, and pick up their leftovers. They also steal food from the other indulgent chimps. Mothers give their babies food, and stop them eating things that are bad for them. One mother at Mahale was seen taking her infant's hand away from some leaves it was about to pick, and when the hand came back and picked them, the mother took them and dropped them. Another mother picked and dropped some poisonous leaves to put them out of her infant's reach. And yet another mother picked and dropped some leaves within her infant's reach, presumably to encourage him/her to eat them.

With such a varied diet it is little wonder that chimpanzees have to learn what to eat, and it is not just a matter of recognizing what is and is not edible, but what is *regarded as* edible, for separate populations do not share the same menu, and only some differences are due to what is available. Of 286 potential food types common to Gombe and Mahale, only 104 are eaten by both groups. They select different kinds of ant, different plants, and even different parts of the same plant. At Gombe, the fruit of the oil-nut palm is the single most important food, and the pith, dried flower stems and dried wood fibres are also eaten. Some West African chimps eat only the kernel, others only the pith, and Mahale chimps ignore it completely.

There are also variations in the way food is collected and prepared, and in the tools used. For example, Gombe chimps bang hard-shelled fruit against tree trunks or rocks, but Mahale chimps use their teeth. Some termite-fishers use both ends of the probe, others only use one end. Drinking with leaf sponges is common practice at Gombe but has only been seen once at Mahale, and only West African chimps use hammers and anvils.

These traditions are passed from generation to generation. Young chimps learn from example. They watch how things are done and then put in a lot of practice. A few bits of grass squashed together, for example, are an infant's first attempts at nest-building. From about one year old onwards, young chimps make day nests far more frequently than adults do – clearly practising for the time when they have to make and sleep in their own beds at 5-6 years old.

Twigs, pieces of bark, and other objects are picked up, pulled apart, and carried around from the age of about 8 months. That a baby can already perceive the potential usefulness of objects demonstrates its considerable brainpower: taking notice of the thing in the first place, bothering to pick it up, deciding what to do with it. Having watched its mother, a 2-3-year- old makes its first stab at a termite mound but will not get its first taste of success for another few years. Some tools are easier to make and handle than others: 5-year-olds are proficient leaf spongers but it takes the whole of childhood to master termite-fishing and nut-cracking.

By about 3 years old the youngster can recognize and help itself to all the 'easy' foods that just need picking.

Giving them bits of nut must also encourage them to stay around and

Youngster watching its mother tool-using (at a termite mound).

watch how nuts are cracked, says Christophe Boesch, who has observed their years of trial and error. For a long time it was thought that no animals actively taught their offspring but one of the Tai chimpanzees gave Dr Boesch a clear demonstration of her teaching abilities. He describes how Ricci watched her daughter Nina struggling for nearly ten minutes with a hammer that she was holding wrongly. 'Finally,' he writes, 'Ricci got up and walked over to her daughter, who immediately handed over the troublesome hammer. Very deliberately, Ricci slowly rotated it into the best position for nut-cracking. As if to emphasise the significance of this movement, Ricci took a full minute to perform the simple rotation. Then, with Nina sitting in front of her and watching carefully, Ricci opened 10 nuts (of which Nina ate 6 whole ones, and part of the rest) and then left her daughter to carry on. Nina resumed cracking nuts, adopting exactly the same grip on the hammer her mother had demonstrated, and successfully opened quite a few.'

In the process of practising tried and tested techniques, young chimps are likely to stumble across new ways of doing things. Watching a youngster who kept dipping a grass stem into a puddle and sucking the drops from it, Jane Goodall points out that this sort of experimenting probably led to the invention of the leaf sponge. She has also seen a youngster squash an insect with a stone, and wonders if the Gombe chimps will one day discover nut-cracking!

Hand-in-hand with practical skills goes social knowledge. Once a youngster knows who's who it must begin to find its way round all the relationships in the group. Once it understands what is going on, it starts being able to predict what is going to happen next and to adjust its own behaviour accordingly. Infants soon learn that it is a good idea to stay close to their mothers when two chimps are getting edgy and an argument is brewing. And they begin to see how the hierarchy works: one particular chimp, for instance, is far more likely to let you

Chimp playing.

take his or her food if your mother is nearby to back you up, but if that chimp's ally is also in the vicinity you have to decide whether your mother dominates the two chimps combined. Through this kind of assessment the youngster learns the consequences of its own actions and begins to see how it can play an active part in the course of events.

The ins and outs of grooming are also picked up over the years. Babies as young as 2-3 months start going through the motions of grooming themselves and their mothers but they cannot do it properly until they are about 2 years old. At about 6 months they start to solicit grooming from their mothers and by about 4 years old they are well aware of the give and take, soliciting from other chimps and responding to requests.

Play also teaches give and take. Between the ages of 2 and 4 years old, chimpanzees are at their most playful. Mothers and siblings are their most common playmates – and peers when families meet up. As well as giving them the chance to practise running and climbing, and sharpening their reactions, games help to teach chimps how to mix. During a tug of war with a stick or a chase round a tree stump, chimps pit their wits against each other and learn to take account of each other's moods and feelings. Adult chimpanzees show a great deal of concern for each other, and playing helps youngsters to become less

Weaning: mother with juvenile having a tantrum.

juvenile showing a pout face.

self-centred. They are upset if a playmate gets hurt or stuck up a tree, and they also find out what is acceptable behaviour and what is not – for games have rules. To keep on good terms, self-control is needed. Someone who bites too hard or gets too rough is likely to be grabbed or bitten back. And while a very young chimp gets away with blue murder, a 3-year-old receives a glare or a raised arm threat for jumping on the head of a sleeping adult.

But the really rude awakening comes with weaning. When her child is about 3½ years old, a mother starts to detach herself, stopping it from suckling and refusing to carry it. She starts gently, playing with or grooming it at the same time as using one arm to block access to her nipple. At first she gives in at the slightest whimper but then hardens her heart and turns deaf ears on the screams of protest that can explode into temper tantrums. Wailing and flinging itself about, the baby seems totally beside itself, but sometimes it pauses mid-scream to see if it is having the desired effect. Some youngsters employ even more devious means, screaming as if they are hurt in order to bring their mothers running to console them. In general, males rant and rage more, and females try to worm their way into their mother's affections or wear her down with whingeing. Weaning does make them very unhappy. They play less and take less interest in everything, and their feelings of rejection are summed up in the way they sit hunched and whimpering.

As if to rub it in that their mother's world has stopped revolving round them, she resumes her sexual swellings sometime during the year-long weaning period so her attention turns to the adult males. Infants often hit and bite their mother's sexual partner who does his

Juvenile playing with a baboon.

best to ignore the interference. Whether they like it or not, watching sex is part of their training for adulthood; they also practise sex from an early age, and a two-year-old male can achieve full intromission of an adult female – most of whom tolerate these attentions.

Whatever it feels like, the infant has not been abandoned. Whenever a mother refuses a request she gives a pat or hug of reassurance, and grooms much more frequently than before weaning began. All chimps have a deep-rooted need for physical contact, and grooming replaces suckling and being carried.

Some mothers do not fully wean their child until the next baby is born but others manage to get a breather for a few months or even years. The birth interval can be anything from 4-8 years. When the new baby appears the mother takes care not to exclude her older child, grooming it more than she grooms the baby, and allowing it to get closer to its sibling than anyone else. Some mothers even try to carry both at once, and stagger about with the big one riding piggyback and the little one slung underneath.

By the time it is weaned the young chimp stands on its own two feet, makes its own nest and sleeps alone, collects most of its own food, and is what humans define as a juvenile. Jane Goodall sums up the physical separation: 'One important distinction between infants and juveniles is that the latter can stand a period of long, heavy rain sitting alone without being hugged by the mother.'

Juveniles find themselves not just kept at arm's length, but also left behind. Until now they have been fetched and carried, and have felt their mother's hand yanking them along whenever they have strayed away from her. Now they must learn to keep an eye on her, and not panic if she does disappear for a second.

The next big step is to take the initiative and actually *choose* to be away from her. The bond between mother and child is still strong, but over the next ten years the relationship slowly shifts from dependence to mutual support. And over those years the child continues to develop its social and practical skills.

Puberty marks the start of adolescence at about 8 years old. Females have their first tiny swellings then, and their first full one when they are about 10 years old, but they do not conceive until several years later. Males first ejaculate at 9-10 years old, but are not fertile until 13 or 14 years old. Females are fully grown at about 19 years old, males at 16 or 17.

Sex differences in their behaviour also emerge. A young female can find everything she needs within her immediate family. She forms close

Youngster practising nest-building.

relationships with her siblings, and helps to care for the younger ones. By early adolescence she is moving on to a more equal footing with her mother, backing her up in arguments, grooming her frequently, and caring for her. When one female at Gombe sat at the foot of a tree because she was too old and sick to climb and forage, her adolescent daughter came down and brought her food.

Both females and males play with their mothers and siblings, but juvenile males also seek each others' company. Trials of strength, assertiveness training and male bonding all feature in their rough games, helping prepare them for entry into adult male society. A young male can be big and tough one minute and sheltering under his mother's wing the next. But although he spends a lot of time with her, his mother is not the magnet she once was. His attention is drawn to the adult males. He watches them, imitates their displays, edges towards them, and eventually starts to follow them around.

Unlike babies who bounce on and off big males without a second thought, juveniles and adolescents treat them with respect. Keen to pay homage and be noticed, young males zealously present and pant-grunt to their superiors. Adult males usually take the hero-worship in their stride but cannot help getting annoyed when an over-demonstrative adolescent actually blocks their path or interrupts what they are doing. A glare or a raised arm threat soon gets rid of him.

But not for long. The young male hangs around the adult males more and more, and sometimes singles out one to attach himself to. The chosen one – often an elder – seems quite willing to take on his apprentice, even waiting for him when the party is on the move.

Travelling with older males gets a young chimp in on some real action such as hunting and boundary patrolling. Mostly he is left to pick things up as he goes along, but sometimes he has to be taught a lesson. One adolescent who would not keep his mouth shut during a boundary patrol received a wallop. When this did not work he was given a hug, which did.

At the same time as working his way into male society, the adolescent starts throwing his weight around with the adult females, charging at them and disrupting their feeding or resting. At first they may be reluctant to acknowledge his new stature, and get together to fend him off, but by the time he reaches adulthood he will dominate them all. Once he has asserted himself, there is no need to be so annoying: adult males receive respect from adult females without having to demand it. Even though an adult male is officially higher ranking than his mother, he retains his respect for her and often travels with her unless she is in oestrous.

It seems that familiarity breeds incest avoidance, for mothers and sons hardly ever show sexual interest in each other, and when females stay in their own community, brothers and sisters rarely mate. Since the identity of fathers is unknown they probably mate with their daughters, although some young females resist older males – who are, anyway, far more attracted to the transferred females they do not know.

A male is regarded as an adult when he has inched his way to the heart of the community and become a regular participant in male-only activities including their exclusive grooming sessions. As a full member of the club he now begins to climb the social ladder. Many females, on the other hand, embark on their adult lives by leaving the community and taking their first steps towards becoming mothers and forming close bonds with their own children.

MOTHERLESS CHILDREN

The emotional bond between them is so strong that when a mother dies, her child often dies too. Orphans under 2 need their mother's milk, but many older ones cannot bear the loss and simply give up, even when their older brothers and sisters look after them.

The first orphan known at Gombe was Merlin, whose mother died when he was 4-5 years old. His elder sister adopted him, carrying him, comforting him, and sharing her nest with him, but even so he grew listless and withdrawn, and died 18 months later. Even one 8½-year-old male could not get over losing his mother. At first he stayed near her body, then became depressed, and died only 3½ weeks after her.

Orphans have much more chance of surviving if they are adopted. Older juveniles and adolescents who lose their mothers and have no immediate family tend to form a close friendship with an adult of the same sex as themselves.

OPPOSITE: *Baby for sale in an African market.*

• 6 •
Out of Africa

T hat baby playing at the start of Chapter 5 has a price on her head. Someone the other side of the world is prepared to pay up to $25,0000 (nearly £14,000) for her. Instead of tumbling into her brother's lap, she could find herself grabbed, bound hand and foot, and stuffed into a sack to be sent out of Africa to feed the 'First World' demand for baby chimpanzees as pets, showbiz gimmicks and research tools.

Babies are wanted because they are so much easier to handle than adults. Hunters are expert animal trackers and creep up on unsuspecting groups of chimps. But the baby's mother and brother will not let her go without a fight, so they will be killed along with any others who get in the way. They may be shot messily with ancient flintlocks loaded with bits of metal that spray everywhere (babies clinging to their mothers end up dead or wounded too), or they may be caught in wire snares, or be set upon by a pack of dogs. Some will be sold as meat. In parts of Africa, including Ghana, Liberia and Zaire, chimpanzees are still a major source of protein – so money can be made from dead adults and live babies.

In her sack, box or basket, the baby will be taken to an animal dealer who probably lives in a town several days' journey away by bumpy truck. Chimpanzee babies need their mothers as much as human babies do. They need their milk and their comfort. On this first leg of their journey many die from dehydration and starvation, from wounds that turn septic, from being tied up so tightly that their circulation stops. Shock and depression also kill. The survivors are handed over to the dealers who sell some as pets in local markets but pack most off to airports and ports from where they will travel across the world to places such as the USA, Japan and Europe. More die in dingy holding compounds, stored like any other merchandise awaiting dispatch, and still more die on their long journeys. Many an importer collects his purchases only to find he has bought a box of dead bodies and is put to the bother of sending for more.

His order seldom goes through official channels. One reason for so much death and suffering is that most of the trade in chimpanzees is illegal, and the black market has no rules. A smuggler's main concern is to hide his contraband, whether it be drugs, weapons or live animals.

On paper, chimpanzees are well protected. The buying and selling of wild-caught animals is controlled by the Convention on International Trade in Endangered Species (CITES). The more endangered the species, the stricter the controls on its commercial exploitation, and chimpanzees are up there with African elephants and giant pandas as one of the most endangered, best protected species in the world. Even so, you can legally take a chimp from the wild providing you can convince CITES officials you have a very good reason for doing so. Most successful applicants are zoos, and permission comes in the form of special export and import licences.

But CITES is not a law, it is a treaty, and the 115 member countries vary in how seriously they take it. To give it any teeth at all each country has to bring in its own laws. Many do. Across Africa there are laws prohibiting the killing, capturing and exporting of wild chimps; across the world there are laws banning their importation. But the day-to-day enforcement of those laws means time, money and manpower. In Africa it means patrols to keep people out of protected areas, to search vehicles and catch hunters and dealers red-handed. Outside Africa it means playing cat and mouse with smugglers determined to ply their lucrative wares – for little chimps are big money. A poacher can sell one chimpanzee for $50-$400 (£27-£222), which probably equals many months' wages. A dealer abroad will pay his supplier in Africa $10,000-$15,000 (£5,500-£8,333) for one chimpanzee, and then sell it for twice as much. These people do not mind paying the odd fine – the most likely punishment for getting caught. And they can spare the backhanders that help officials look the other way.

Most chimpanzees leave Africa as stowaways in tiny boxes, but some sail through customs in full view, even poking their heads out of people's hand luggage. Sometimes the red tape is cut, other times it is just not there. Only 12 of the 1,026 chimpanzees exported from Sierra Leone in one 5-year period were registered with CITES even though most went to countries that were members. Some slip through with forged papers. It is legal to buy and sell captive-bred chimpanzees, so many wild-caught ones are simply registered as captive-bred. Some dealers invent their own non-existent zoos, whose official-looking notepaper give them perfect cover. Or else the chimps are laundered: sent to a non-CITES country or one with the most lax laws, and re-exported as captive-bred from there.

How many chimpanzees are living in laboratories, zoos, circuses and private homes is unknown, but it is thought to be as many as 6,000, most of whom were taken from the wild. It is estimated that 1,000 chimpanzees are captured every year. Add to that the 10-30 that die for every one that survives its first year in captivity, and you can see how trade is helping habitat destruction to push chimpanzees towards extinction. And they often go in tandem: clearing the forest makes it easy for hunters to move in and snap up besieged chimps – and removing mothers and babies wipes out the next generation.

Back to that baby. Supposing she has not joined her mother and brother in the 10-30 that die; supposing she is the one that survives. She might find herself in a holiday resort in Spain or the Canary Islands. There she will be thrust into the arms of tourists who each pay £5-£8 for the pleasure of having their photograph taken with this living cuddly toy. The children's clothes she is squeezed into will make her hair fall out and give her skin problems. The shoes deform her feet. She may have her front teeth pulled out to stop her biting. She may be drugged to keep her quiet and floppy. She may be beaten, and stabbed in the face with a lighted cigarette.

She will work non-stop for up to 16 hours, with several different photographers: beaches by day, discos by night. She is unlikely to survive more than a few months. Those that do are discarded when they start losing their babyhood at the grand old age of four years old. They are sold to anyone who will take them, or just abandoned. One was found in a dustbin with his throat slit. Others have been found drowned.

Spain signed CITES in 1986, at the same time as joining the EEC, but separate districts have their own laws, and year after year the authorities continued not to notice the chimpanzees even when people reported exact times and places. In 1991 the EEC almost took Spain to the European Court for continuing commercially to exploit an endangered species, but was persuaded to drop the charges when it was assured by top government officials in Spain that they would be dealing with the matter. It seems that Spain is now acting, but even if these photographers' dummies do, at last, disappear from Spain, they are beginning to pop up in Israel and Mexico.

But being a photographer's dummy is not the only way the baby chimp may be used to entertain us. Dressed, she may roller skate round a Las Vegas nightclub kissing hands and taking sips of wine. If she steps out of line she receives a jolt of electricity from the little box on her back worked by remote control. Or she may be shown a rolled-up newspaper inside which is the electrified cattle prod that has been used to train her. All over the world there are chimpanzees performing in circuses and magic shows, or put on display in shops and hotels to attract the customers.

Pushing the 'aaahhh' button in all of us, the baby chimpanzee may become a pet, bought for her novelty value or because someone felt sorry for her when they saw her in a shop. But chimpanzees are not made to be pets. She can play the role for a while, can be sat at table, taught to use a knife and fork. There are chimpanzee pets who take the helms of their masters' yachts, chimpanzee pets with their own pet cats. She will make you feel good and needed. Without her mother and brother she will turn to you, cling to you, cry when you go.

But sooner or later she will open the fridge and all the cupboards, turn keys in their locks, climb the curtains. She will start to grow into a big

Spanish beach chimp.

strong animal with a mind of her own. Half-human, half-chimpanzee, she will be mixed up, pent up and dangerous. She will be 'put to sleep', or condemned to a cage for life, or sold to a zoo or research laboratory.

Whether ex-pets or fresh from Africa, most wild-caught chimpanzees end up in laboratories. The baby could be sealed inside a metal box for the rest of her life. The box will be small and bare; she will have nothing to do, and no company for the rest of her life. Jane Goodall writes: 'I watched one of these older chimpanzees, a juvenile female, as she rocked from side to side, sealed off from the outside world inside her metal isolation chamber. She was in semi-darkness. All she could hear was the incessant roar of air rushing through vents into her cell. When she was lifted out by one of the technicians, she sat in his arms like a rag doll, listless,

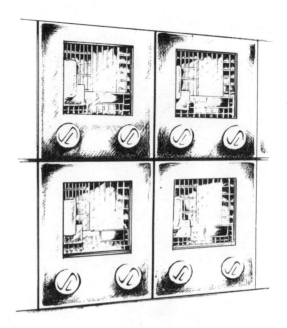

Metal laboratory cages.

apathetic. I shall be haunted forever by her eyes, and by the eyes of the other chimpanzees I saw that day. They were dull, blank, utterly without hope.'

The scientists say the conditions are necessary for their experiments. If she was with other chimpanzees she might give them whatever disease she has been infected with, and ruin the research. While they wait for the injected disease to develop inside her, she cannot have any bedding or toys in case she picks up something nasty from them. Also, an uncluttered cage is much easier to clean, and a small one makes it much easier to catch her.

About 4,000 chimpanzees live in laboratories. They are used mainly in hepatitis and AIDS research. It is estimated that 300 million people carry the virus for hepatitis B. By using chimps, scientists have developed a vaccine against the virus.

Now the race is on to find the cure for AIDS. Fully aware of the kudos and the money that will go to whoever gets there first, the drugs companies are not just racing against time but against each other. So far, chimpanzees and gibbons are the only animals that have been successfully infected with HIV isolated from human AIDS patients. But so far, they have failed to become very ill, so that are not as useful as the researchers were hoping. They keep trying though, and it may take a long time, a lot of blind alleys, and a lot of chimpanzees before they come up with anything. And if the chimps do what they are supposed to they will be much in demand for testing the new drugs. Today, at least some hepatitis B vaccines are made from genetically engineered virus parts that require no chimpanzee testing, but by law all the first batches of vaccine had to be tried out on chimps before being passed fit for human consumption.

People will argue till the cows come home whether we have the right to carry out any experiments on any animals, or where we should draw the line: fly? fish? frog? rat? rabbit? chimpanzee? With chimpanzees the arguments are circular – they should not be used for experiments for exactly the same reason that the scientists say they must be used: that they are so like us. But it is not just their livers and kidneys that are so like ours, it is their need for each other, their lively minds, their feelings. Believing that they are justified in doing to a chimpanzee what they would not dream of doing to a person, the experimenters seem to have immunized themselves against what they are dealing with. Jane Goodall has persuaded some labs to make their chimps' lives more bearable by keeping them in larger, less uncomfortable cages. Some are even kept in groups and given toys to play with. Goodall and other chimpanzee experts are thankful for all the small mercies they can get for those 4,000 individuals.

But they also want to save wild chimpanzees from this fate. In the past, whenever their stocks ran low the labs simply ordered more from Africa, but CITES makes it much more difficult to get supplies, so they have been forced to look into producing their own. So far there is a lot of talk about breeding all they need, but the chimpanzees have not been very co-operative.

The message does not seem to have got through to those who have spent their lives in metal boxes that they must wake up and play mummies and daddies. During the 1980s, the US biomedical establishment spent more than $10 million in the first 4 years of setting up special breeding groups, but their target numbers seem way beyond reach.

Wild-caught chimpanzees continue to find their way into labs, some with and some without CITES blessings. Although talk of 'cropping' from the wild has now dropped out of fashion, some scientists think that a shortage of chimpanzees could hold back progress in AIDS research, and some are showing a keen interest in the African countries with the healthiest numbers, even volunteering to fund their own surveys of wild populations.

Some drugs companies have added a new twist to the tale: if the chimpanzees cannot come to them, they will go to the chimps. Building laboratories in Africa bypasses the laws, gets you out of the public eye, and guarantees you a steady supply of chimpanzees. There are already research stations in Liberia, Gabon and Zaire, and there is talk of establishing them in Uganda, Ivory Coast and Nigeria.

Wherever they are, chimpanzees experiments are carried out behind closed doors; it is easy not to think about them. The chimpanzees on show – the clowns and zoo inmates – have always been there; it is easy to see them as completely different animals from the awe-inspiring wild ones of the African forests. When you do think about the varying degrees of pain and impoverishment suffered by 6,000 or so captive individuals – plus the 1,000 or so being captured in the wild every year – it is easy to feel powerless. But a handful of people are showing what a difference a handful of people can make.

CHIMPS IN COURT

Chimpanzees have appeared in court several times in recent years, some literally – held up as living proof of smugglers' activities – but most represented by people, some of whom have had multi-million dollar libel suits slapped on them for speaking up on the chimps' behalf.

Some of the biggest legal battles have concerned an Austrian multinational drugs company, Immuno AG, which carries out research on chimpanzees. Immuno was thrust into the limelight in the early 1980s when an Austrian animal protection society accused the company of cruelty to chimpanzees because its housing conditions were so poor. At the same time it was in the news for ignoring international trade regulations by trying to import wild-caught chimpanzees that were confiscated by the Austrian authorities.

Soon afterwards, it became known that Immuno intended to set up a branch in Sierra Leone. When various conservationists, primatologists and journalists expressed in writing that this could be a very handy way of cutting all the red tape, each one was sued $4 million for libel. Eventually all but one settled out of court, and that one fought and won the case by defending his freedom to express an opinion.

Immuno's plans did not get very far in Sierra Leone because it is illegal to take any chimpanzees from the wild. But it used to be allowed, and, together with habitat destruction, has reduced the numbers from 20,000 to 2,000. One major dealer is Dr Franz Sitter, an Austrian expatriate who has been exporting chimps from Sierra Leone for nearly 30 years. Last time he totted it up, he reckoned he had sold about 2,000 chimpanzees to the US alone. It is thought that nearly three-quarters of the world's captive chimpanzees passed through his hands. (Don't forget that 10-30 die for every one that reaches its destination.)

Immuno is one of his customers. When it failed to set up in Sierra Leone, it bought 20 chimps from Sitter in 1986, and imported them to Austria. Sierra Leone has banned all exports of chimps since 1979, but Sitter used an old permit and this was accepted by Austria – thus starting another round of legal wrangling and a public outcry on behalf of the chimps; they are at Immuno, and when last heard of were earmarked for AIDS research in a joint venture with US researchers.

• 7 •
Born Free?

On an island in the River Gambia, West Africa, lives a group of over thirty chimpanzees. Like normal chimpanzees, they spend their days eating and travelling, grooming each other, playing, squabbling and lazing in the shade. But these are not normal chimps. They have been rescued from human hands, and are members of the Chimpanzee Rehabilitation Project. They have had to learn how to live in the wild again.

The project began in the early 1970s when government wildlife official Eddie Brewer confiscated young chimps being smuggled through the Gambia to be shipped out of Africa. With the help of his daughter Stella, he took care of the orphans and gave them a taste of the wild by taking them for walks in the bushland of the national park where the Brewers lived. Watching the chimps in the trees made Eddie and Stella want to return them to the wild – but there is a lot more to granting chimpanzees their freedom than leading them into the wilderness and waving good-bye. Chimps that have been wrenched from their natural way of life need a lot of time and understanding if they are to fit back in. They have missed out on the childhood that teaches them how to be chimps, and the support and security of their fellow kind has been replaced by brutality and neglect. Each member of the project bears physical and mental scars. As well as those confiscated in transit, there are some that have served time with the photographers in Spain, or as pets, experimental tools or zoo inmates. The scars do not vanish overnight, but they do fade.

It can take years. The remote, forested island in the river is an ideal halfway house for the chimps because project staff live in a camp on the mainland and cross the river by boat, so they can control how much time they spend with the chimps. The rehabilitators need to support and reassure the chimps enough to help them develop the confidence to move *away* from people, not closer to them. It is all too easy to make an orphaned chimpanzee need you.

During the earliest stages of rehabilitation the chimpanzees were accompanied for walks every day by people who uttered excited food grunts at the sight of every edible plant, who chewed leathery leaves and bitter fruits with gusto, who climbed trees, built nests, and ran away from snakes. In a rather strange version of teaching grandmothers what to do with their eggs, humans imitated chimpanzees to help them get used to the forest and to each other.

At the same time as bringing out the chimps' natural behaviour, these teachers helped them forget what other people had taught them. Some, for example, were not only expert pickpockets and undoers of shoelaces, but could readily unscrew a bottle top, pour themselves a drink and hold the

'...absorbed in scribbling with a pencil on some paper'.

cup without spilling it – but had no idea how to go about opening a hard-shelled fruit.

Orphans who were together in captivity are often inseparable, but others have not seen another chimp since the day they were captured, and would much prefer to sit in your lap all day. In time even the most dubious introvert starts to join in, and the human minder can gradually withdraw. Some new recruits, for instance, consider only people fit playmates, so whenever another chimpanzee sidles up during a game, the human plays with both at the same time, and before they know it they are playing with each other.

It does not take long for most of the chimpanzees to form close friendships. It is quite common to see a youngster (who should still be taking rides on his or her mother) struggling to carry a smaller or less confident companion, and some of the older females have adopted young ones. All of them have to become old before their time. Without their mothers and other adults, they take on adult-like roles – someone is the best forager, someone else the best peacemaker.

Much is being asked of them, but out of this motley collection of traumatized individuals has grown a close-knit community well able to look after itself in the bush. You could not expect them to forget their pasts completely, and every so often they remind you that they are straddling two worlds: a juvenile, one of the best at finding food, is absorbed in fishing for termites. Like any wild chimpanzee of his age, he is not much good at it, and his stick is far too short and thick, but he keeps trying. The next day that same chimp sits in a clearing in the middle of the African forest, absorbed in scribbling with a pencil on some paper. Scribbling must have been a favourite pastime in his previous life as a pet, and one of his minders has, just this once, given in to his begging to have a go. You can also guess which ones used to be beach chimps: most chimpanzees dislike looking at a camera with its big, predator-like eye, but whenever they see a camera pointed at them, the ones from Spain stop what they are doing and pose.

Apart from these revealing details, the chimpanzees are living a wild life, and many of the females have now had babies. Some lost their firstborn because they did not know how to look after them, but now they learn from each other, and humans play little part in their lives. But how well the chimpanzees cope with the wild is only one side of the rehabilitation coin. The other is how well 'the wild' copes with them. The island was not made for chimpanzees. They have disrupted its delicate ecology, and compete with wild monkeys and other animals for food. Even though they are capable foragers, there is not enough food to keep them going, so they receive regular handouts of mangoes, grapefruit, oranges and peanuts.

The original plan was to move rehabilitated groups off the island so the next batch could take their place. But there is nowhere for them to go. This might sound odd when you think how large Africa is, but chimpanzee habitat is shrinking fast. Also, rehabilitated chimps have a long list of requirements before an area can be considered suitable for them – one of which is that there should be no wild chimps. Some of the first rehabilitants were released in a national park in Senegal but were attacked by wild chimpanzees, so they were moved to the Gambian island. Even in the unlikely event of the newcomers being tolerated, they could overcrowd the area, causing stress to the chimps and the habitat.

Any unoccupied suitable-looking spots need careful investigation. If chimps have never been in residence it is probably not as suitable as it looks, and if they have all died out there must be a reason. There is little point in handing rescued chimps back to the hunters, or giving them a home already on the demolition list. Even if the land is officially protected, nowhere is 100% poacher-proof, and humanized chimpanzees would be easy prey since they have lost the fear that keeps wild ones away from people. Or else they could be dangerous, and attack intruding humans instead of fleeing from them.

It looks as if the rehabilitated chimpanzees will be staying on the island,

Youngster investigating.

which seems to be fine by them, but means no more can be put there. It also means the project must run indefinitely to feed and care for them and their children and – presumably – their children's children.

The project grew out of a desire to give the chimps a second chance – which is also how the best chimpanzee sanctuary in the world evolved. Ten years ago, David and Sheila Siddle had never had anything to with a chimpanzee. Now they have taken in 35.

Having worked in Africa most of their lives, this British couple bought some land in an isolated corner of Zambia, turned it into a cattle farm, and settled down to a quiet retirement. But Sheila has a reputation for looking after waifs and strays. Jackal puppies, cane rats, a baby hippo and a fruit bat are just some of the animals she has taken under her wing, and in 1983 wildlife officials turned up with a baby chimpanzee.

Sick and starving, with his teeth smashed in and his face gashed open, he had almost given up. It took four months of round-the-clock nursing before he was well again. Word soon spread that at last there was some-one not only willing but *able* to look after rescued chimpanzees, and officials were able to start clamping down. There are no wild chimpanzees in Zambia, but hunters were catching them in neighbouring Zaire and smuggling them across the border to be exported or sold locally as pets.

Time and again the Siddles have dug out bullets, treated dehydration, dysentery and malnutrition, and saved lives. Chimfunshi Cattle Ranch (Chimfunshi has nothing to do with chimps, but is a tributary of the Kafue river) was turning into Chimfunshi Wildlife Orphanage. Like the Brewers, they took the chimps for daily walks in the bush; knowing they could not just release them, and rejecting any thoughts of their being caged, David and Sheila decided to do things the other way round, and enclose a piece of the bush by building a brick wall round seven acres of it. With little money and no machinery, they and their farm workers started from scratch, digging up the earth and making the bricks. Going up at the same time as the Berlin Wall was coming down, it took two years to build, and has been dubbed the Great Wall of Zambia.

Behind the wall live sixteen rescued chimpanzees who are fed and watched over and treated when sick, but otherwise left to their own devices – to forage, climb trees, make nests and generally live like wild chimpanzees. Two babies have recently been born. Thanks to care and understanding, and their own remarkable resilience, the chimps, like the ones in the Gambia, have sorted themselves into a very well balanced community.

More keep turning up at the Siddles' door. They include a male and female abandoned in New Guinea by a travelling circus, two brothers born in a zoo in Israel, and six under-twos confiscated in Africa. One of these had been smuggled out of Uganda to the United Arab Emirates but was refused entry as there were no papers, and returned to Uganda. When the box was opened at Kampala Airport it was discovered to contain not one but two chimps. The second one was squashed underneath the first, and completely unable to move. They were named Pan and Dora, and for

The horrors of smuggled chimp babies.

the first few months Pan refused to let go of Dora who mothered him even though she was only a baby herself. Then there was Goblin, who was smuggled from Zaire to Uganda where he was impounded. By the time he reached Chimfunshi he was badly malnourished and unable to use his legs because he had been folded into a tiny box for so long. The faces of three baby chimps were spotted peering out of a basket going round the luggage carousel at Nairobi Airport. One died soon afterwards, but the other two, Mike and Grumps, were cared for until they were well enough to travel to Chimfunshi. And last but not least there was Pippa, presented to the Siddles by smugglers who had heard they might be interested in buying chimps. The Siddles said yes they were, but David would have to go to town to collect the money. Instead of going to the bank he contacted wildlife officials and the smugglers ended up spending five months in jail.

Home for these and the other new arrivals will be 14 acres of bushland enclosed by an electric fence – and the Siddles are already making more plans: they have bought a 200-acre piece of land that they want to hand over to all the chimps. A river runs round it in a horseshoe, so that would act as a moat, and they just need to find a way of blocking off the rest . . . 'These people will not stop,' says chimpanzee expert and conservationist Dr Geza Teleki. But the Siddles are rather baffled by all the acclaim they have been getting. They say they are just doing the best they can in the circumstances, and David asks: 'If this is really the best sanctuary in the world, what does the rest of the world think it is doing?'

It is a good question. Jane Goodall is answering it with plans to set up a string of chimpanzee sanctuaries in several countries, including one in Glasgow, on the zoo site: this is in the planning stage at the time of going to press. With branches in the US and UK, the Jane Goodall Institute (JGI) is raising money to help conserve and study wild chimpanzees, and to care for rescued ones. Two sanctuaries are just starting up in Burundi and the Congo. Until now baby chimpanzees on sale in local markets in these countries have been rescued by individual people unable to turn away and leave them. Buying them, they know, feeds the demand, but when you are faced with a dying baby chimp what do you do? Even with the best will in the world some of these people discover they have bitten off more than they can chew – so the ones that *can* manage soon find themselves inundated. In the Congo, for example, over twenty chimps took over Aliette Jamart's house and garden until she released them on an island sanctuary in a river near the Congo-Gabon border.

Likewise, in Spain, Simon and Peggy Templer turned their house and garden into a rescue centre for beach chimps, and battled for nearly twenty years to persuade the authorities to stop the photographers. It was thanks to the Templers that some chimpanzees from Spain ended up in the Gambia, but those more recently rescued went not to Africa, but England. Monkey World in Dorset was set up in 1985 as a refuge for unwanted primates, and its director, Jim Cronin, has pledged to give a home to every single chimpanzee from Spain.

When he was director of National parks and Tourism in Zambia, Harry Chibwela likened the Chimfunshi chimpanzees to refugees. People are doing all they can to save lives, but they know this is only a stop-gap, and that the roots of the problem must be tackled: the trade must be stopped – which means stopping the worldwide demand – and the habitats protected. Chimfunshi and the other African sanctuaries are already showing that they can help cut down the numbers of chimpanzees captured because officials are far more likely to confiscate them when they know there is somewhere for them to go, and word filters down to the hunters that chimpanzee catching is likely to land them behind bars.

Longer-term and more positive than the threat of punishment is to change attitudes to chimps; most of the rescue workers are convinced that once people are given the chance to find out about chimpanzees they will want to help and protect them. Some sanctuaries are open to visitors, and also organize travelling exhibitions that visit local towns and villages. Tourists in Spain are handed leaflets that explain the history of the chimp that poses for photographs. The campaigners ask people not just to refuse to have their pictures taken but also to report where and when they met the photographers. The information is passed on to the police. Over the past few years, both the World Wide Fund For Nature and the World Society for the Protection of Animals have leafletted tourists, but the most consistent advocate on behalf of the beach chimps is the International Primate Protection League (IPPL).

IPPL was founded in 1973 by Dr Shirley McGreal when she saw crateloads of baby monkeys stacked in Bangkok Airport on their way to research laboratories. Now with representatives in 32 countries, IPPL acts for all captive and wild primates everywhere. It works with governments to protect wild primates and their habitats and played a large part in persuading Thailand and India to ban primate exports. But much of its energy goes into stopping the illegal trade by tracking down the dealers. IPPL's undercover investigations have exposed international smuggling rings and helped to imprison or indict several people. What with intrigue, threats and lawsuits, this frontline work sometimes seems to have all the ingredients of a good spy story, but the victims are real.

While IPPL gives all primates a voice, chimpanzees have their own spokespeople. As well as the JGI there is the Committee for the Conservation and Care of Chimpanzees (known as the 4Cs). Its members are all chimpanzee experts who want to save wild chimpanzees from extinction and to improve the lives of captive chimps. With their firsthand knowledge and academic clout, these people are not easily dismissed as soppy animal lovers.

They write reports and recommendations, and meet the politicians and other powers that be. Pushing chimpanzees to the front of the political agenda seems like an uphill slog, but a major breakthrough came in 1990 when the US Government put stricter controls on the importation of wild chimpanzees to the US. The 4Cs is now surveying wild populations and producing an Action Plan that will suggest how best to protect chimpanzees in each of the twenty-one countries where they still live, and it is gathering as much hard evidence as it can about the illegal trade. The trouble is that as soon as one pipeline is exposed and stopped another opens up. This will go on so long as the demand is there.

Gregoire.

Many neglected, mistreated and badly housed chimpanzees need refuge, and they cannot continue relying on the goodwill (and shoestring budgets) of individuals who find it impossible to turn them away. Every country that keeps chimps in captivity needs to face its responsibilities to them. Money and thought needs to be put into giving them a much better deal than most of them have now. We cannot give them back what we have taken away and we cannot set them free, but we can release them from loneliness and boredom and abuse, and give them the freedom to be themselves, to live in dignity and to express as much as possible of their natural behaviour. Jane Goodall wrote of one chimpanzee:

> Never shall I forget my first sight of Gregoire. Was this really a chimpanzee? This was an all but hairless creature whose pale skin was stretched so tightly over the emaciated body that every bone could be studied. Who looked out from a dark dungeon with dull eyes, one of which was half white, blind. Who reached out, for an offered morsel of food with a long, claw-like hand.

'Shimpansee – Gregoire, 1944'. It was written over his cage. I gazed in disbelief. In that bleak, dim, cement-floored cage he had endured his captivity for 46 years. And now, I thought, he is dying.

Gregoire lives in a zoo in Brazzaville in the Congo. Jane Goodall arranged for people to visit him and bring regular supplies of food, and later they formed a support group for him and other animals at the zoo. The JGI then sent a representative who employed and trained a Congolese keeper, Jean Maboto, to look after the primates.

One year after first meeting Gregoire, Dr Goodall returned to find a sleek-coated, bright-eyed chimpanzee sitting on a bed of fresh leafy branches grunting and screaming with enjoyment, while he ate three large croissants.

Milla was born in Cameroon just over twenty years ago. When she was about a year old her mother was shot for meat and she was rescued from a meat market by a British couple who looked after her and then gave her to the Mount Meru Game Sanctuary in Tanzania. She lived as part of the manager's family until she got too big and dangerous, when she was put in a cage near the hotel bar. Cooped up there she grew fat on coke, beer and titbits. There she stayed until 1990, when David and Sheila Siddle took her in. Not knowing what she would do to other chimpanzees, having not met any for so long, the Siddles put her in a cage next to a good natured young male, Sandy. Milla promptly put her arm through the bars and shoved him off his shelf, and it was a good three days before he plucked up courage to get within arm's reach of her again. Sheila Siddle writes:

> Milla then turned to Sandy and put out her right arm, and flicked her hand at him in a gesture that I thought meant 'Go away!' Her hand, when she flicked it, touched Sandy's shoulder and he jumped away with a little cry. They were both looking at each other very intently. Then Sandy moved, very, very slowly, closer to Milla. She again put out her arm and flicked her hand. This time Sandy did not jump away when it touched him, again, on his shoulder. Milla put out her arm a third time, flicked her hand, touched Sandy's shoulder and this time gently took hold of a handful of his fur. Sandy seemed frozen to the spot as Milla moved her hand, in very slow motion, up and down, up and down his left arm. This seemed to go on forever, but in fact it was perhaps two minutes. And then Milla moved her hand on to the top of Sandy's head and very gently pulled him closer to her. There seemed to be no reaction from Sandy, but I could feel the tense atmosphere. Only when she put forward her pouted lips and

kissed him on the top of his head did he seem to relax his rigidly held shoulders. Milla then put out both her arms and pulled Sandy towards the bars, at the same time pushing herself up as close as possible so that their bodies touched for a brief hug.

This was the first contact Milla had had with another chimpanzee for nineteen years.

The only thing we can be sure of is that chimpanzees will keep coming up with surprises. However long and hard we look, we will only ever catch the odd glimpse of them: we find it difficult enough to understand ourselves and what goes on in our own heads, let alone get inside theirs. Yet chimpanzees, our closest relatives, are the animals we have the best chance of getting to know. You cannot fail to meet them halfway, to find their laughs infectious and their tempers frightening, and to recognize the smallest ways in which they express affection for each other – a look, a hand resting on someone's leg.

You can not only meet them on equal terms but also learn from them. When you help rescued chimpanzees get back to being chimpanzees, you copy them – biting them in play and getting bitten back, exchanging hoots of hello and indulging in long grooming sessions. But you also pick up things from them without noticing – slipping into their way of being alert and interested in everything around them in the forest and reading the signs they send each other all the time.

Anthropologists living with other cultures inevitably find themselves thinking about their own way of life and looking at themselves from a new angle. As soon as they answer one set of questions they come up with more. For all the detailed dossiers, we are only just starting to make their acquaintance and, with a lifespan of fifty years, even the Gombe chimps have another twenty years to go before the first complete birth-to-death life story is written.

We still know nothing at all about most chimpanzees. Only a handful of groups have been contacted, and we need basic surveys and head counts – or more precisely, nest counts, since today's chimpanzees are just as wary of people as they were when Garner sat in his cage in the forest.

They are right to keep their distance. One hundred years after Garner, Jane Goodall describes chimpanzees as being 'imprisoned' on islands of forest. Many fieldworkers spend more time trying to save them than study them and have to argue the chimpanzees' case, giving reasons why they should be allocated space and left alone. One good reason is that chimpanzees show us that perching in splendid isolation at the top of the tree,

Big male playing with youngster.

passing judgement on them and everything else, is an uncomfortable place to be. They tap you on the shoulder and remind you that you are not as special as you think. They show you to your place – next to them. They remind you that for five million years, ever since humans took one branch and they took another, they have been living alongside us – arguing and making up, calling to each other through the forest, climbing trees and picking figs and doing 1001 things that we know nothing about.

LUCY AND JANIS

With a vocabulary of 120 symbols, and the ability to string sentences together, Lucy was one of the most advanced chimpanzees at the University of Oklahoma's sign language project. She lived with her teachers, Jane and Maurice Temerlin, as a member of the family. After lessons she would help herself to a gin and tonic, turn on the TV, click the switch to find the right channel, and settle back on the sofa.

Janis was a graduate student at the university, and one of Lucy's teachers. They developed a close friendship, and when Lucy grew up Janis helped the Temerlins with their plans to give her a chance to live her own life. In September 1977, Lucy and Janis left for the Gambia. Lucy was to be returned to the wild, and Janis intended to stay a few weeks to see her settled in. Fifteen years on, Janis is still there.

She stayed because it soon became clear that Lucy was not going to slip into the life of a wild chimpanzee easily. Born in a roadside zoo in the US, she had never set foot in Africa before. She grew thin and lethargic, and after a year she showed little interest in her new lessons: instead of playing word games, Janis was now trying to teach Lucy some survival skills such as what to eat and how to build a nest.

Lucy was not Janis's only pupil. She soon collected a small group of rescued chimps, three confiscated in Africa, plus four that were part of a consignment illegally exported from Sierra Leone and stopped by customs officers in Holland. Being younger than Lucy, they found it easier to adjust and had less to let go. Janis wondered if Lucy was just too humanized and too stuck in her ways to make it.

But very gradually, over a number of years and with Janis's constant help, Lucy took to the forest and the other chimps. Important events for Lucy and Janis included Lucy's adoption of a rescued baby chimp, Marti, and his death 4 years later from a rare allergic reaction to a parasite. Lucy herself nearly died in 1983 from a severe hookworm infestation, and her life was saved by being given two transfusions of blood rushed from the US.

Ten years after Lucy and Janis had arrived in Africa, Lucy was at ease in the forest, and she and Janis began to go their separate ways. Janis writes: 'The withdrawal was difficult for both of us. There was an overlap of our lives that I could not separate easily . . . When we departed for Africa, we both left behind our loved ones, our material possessions, and a comfortable lifestyle. We suffered through many of the same depressions, illnesses and deprivations during our transition. I couldn't tell where my life stopped and hers began.'

At last, Lucy and her group were left to live like wild chimps, but were visited regularly just to make sure everyone was all right. Then Lucy disappeared, and despite searching and searching for her, Janis never saw her again. Her skeleton was found with its hands and feet missing. Janis does not know how she died, but suspects she was shot and skinned. Lucy never learnt to fear people and would not have fled from a hunter.

Janis still lives in the Gambia, and oversees the rehabilitation of other chimpanzees. She also runs a conservation education project and visits local schools and villages to talk about the chimpanzees, and the value of conserving the forests and their wildlife.

Useful Addresses

The Jane Goodall Institute (JGI)
Dilys Vass
15 Clarendon Park
Lymington
Hants SO4X 8AX

The International Primate Protection League (IPPL)
Mr C. Rosen
116 Judd Street
London WC1H 9NS

Monkey World
Jim Cronin
Longthorne
East Stoke
Wareham
Dorset BH20 6HH

Committee for the Conservation and Care of Chimpanzees
3819 48th Street NW
Washington
DC 20016
USA

Further reading

Boesch, C., 'First Hunters Of The Forest', *New Scientist*, 19.5.90, pp.38-41.

Boesch, C. & H. Boesch-Achermann, 'Adventures In Eating', *BBC Wildlife*, Oct 1990, pp.668-672.

Boesch, C. & H. Boesch-Acherman, 'Dim Forest, Bright Chimps', *Natural History*, Sept 1991, pp.50-56.

Boesch, C. & H. Boesch-Achermann, 'Forest Close-Ups', *BBC Wildlife*, Jan 1992, pp.14-20.

Desmond, A., *The Ape's Reflexion* (Quartet, London, 1980): discussion of ape language experiments. Dated, but basic questions still relevant.

Diamond, J., *The Rise And Fall Of The Third Chimpanzee* (Vintage, London, 1992): puts us in our place as the third chimpanzee, and then explores what makes us so different.

Ghiglieri, M., *The Chimpanzees Of Kibale Forest* (Columbia University Press, New York, 1984): firsthand account of fieldwork in Uganda.

Goodall, J., *In The Shadow Of Man* (Collins, London, 1971): a classic.

Goodall, J., *The Chimpanzees Of Gombe – Patterns Of Behaviour* (The Belknap Press of Harvard University Press, 1986): full account of 25 years' fieldwork, complete with maps, graphs, charts, and an extensive references list.

Goodall, J., *Through A Window* (Weidenfeld & Nicolson, London, 1990): sequel to *In The Shadow Of Man.*

Heltne, P. & L. Marquardt (eds), *Understanding Chimpanzees* (Harvard University Press, 1989): collected (highly readable) papers from an international symposium, covering recent research on chimps and bonobos, wild and captive.

Kano, T., *The Last Ape* (Stanford University Press, 1992): firsthand account of many years spent bonobo-watching.

Macdonald, D. (ed), *The Encyclopaedia Of Mammals* (Unwin Hyman, London, 1989): good section on primate biology and behaviour.

Nishida, T. (ed), *The Chimpanzees Of The Mahale Mountains* (Tokyo University Press, 1990): fieldworkers' accounts and discussions of behaviour.

Preston-Mafham, R. & K., *Primates Of The World* (Blandford, London, 1992): photo-packed introduction to primates. Includes chapters on reproduction and parental care, social behaviour, habitat and ecology.

Teleki, G., 'Chimpanzee Subsistence Technology: Materials and Skills', *Journal of Human Evolution*, 3, 1974

Tuttle, R., *Apes Of The World* (Noyes Publications, New Jersey, 1986): synthesis of research on apes, including chapters on communication, feeding, and brains and mentality. 70-page references list.

Index

page references in bold indicate illustrations

If you have enjoyed this book, you might be interested to know about other Whittet natural history titles:

BADGERS
by Michael Clark
with illustrations by the author

BATS
by Phil Richardson
with illustrations by Guy Troughton

DEER
by Norma Chapman
with illustrations by Diana Brown

EAGLES
by John A. Love
with illustrations by the author

FALCONS
by Andrew Village
with illustrations by Darren Rees

FROGS AND TOADS
by Trevor Beebee
with illustrations by Guy Troughton

GARDEN CREEPY-CRAWLIES
by Michael Chinery
with illustrations by Guy Troughton

HEDGEHOGS
by Pat Morris
with illustrations by Guy Troughton

MICE AND VOLES
by John Flowerdew
with illustrations by Steven Kirk

OTTERS
by Paul Chanin
with illustrations by Guy Troughton

OWLS
by Chris Mead
with illustrations by Guy Troughton

POND LIFE
by Trevor Beebee
with illustrations by Phil Egerton

PONIES IN THE WILD
by Elaine Gill
with illustrations by Diana E. Brown

PUFFINS
by Kenny Taylor
with illustrations by John Cox

RABBITS AND HARES
by Anne McBride
with illustrations by Guy Troughton

ROBINS
by Chris Mead
with illustrations by Kevin Baker

SEALS
by Sheila Anderson
with illustrations by Guy Troughton

SNAKES AND LIZARDS
by Tom Langton
with illustrations by Denys Ovenden

SPIDERS
by Michael Chinery
with illustrations by Sophie Allington

SQUIRRELS
by Jessica Holm
with illustrations by Guy Troughton

STOATS AND WEASELS
by Paddy Sleeman
with illustrations by Guy Troughton

URBAN FOXES
by Stephen Harris
with illustrations by Guy Troughton

WHALES
by Peter Evans
with illustrations by Euan Dunn

WILDCATS
by Mike Tomkies
with illustrations by Denys Ovenden

Each title is priced at £7.99 at time of going to press. If you wish to order a copy or copies, please send a cheque, adding £1 for post and packing, to Whittet Books Ltd, 18 Anley Road, London W14 0BY. For a free catalogue, send s.a.e. to this address.